D1289046

BLAZING FOREST TRAILS

CHARLES D. SIMPSON

He served an apprenticeship as a fire guard when the Forest Service was only six years old. Since then he has filled many posts, including that of assistant regional forester and supervisor of four national forests in three different regions. He is now retired. He helped, figuratively and literally, to blaze many Forest Service trails.

BLAZING
FOREST
TRAILS

By

CHARLES D. SIMPSON

and

E. R. JACKMAN

ILLUSTRATED WITH PHOTOGRAPHS

THE CAXTON PRINTERS, Ltd.

CALDWELL, IDAHO

1967

© 1967 BY
THE CAXTON PRINTERS, LTD.
CALDWELL, IDAHO

Library of Congress Catalog Card No. 67-12493

Printed and bound in the United States of America by
The CAXTON PRINTERS, Ltd.
Caldwell, Idaho 83605
105889

CONTENTS

ILLUSTRATIONS

BLAZING FOREST TRAILS

CHAPTER ONE

THINKING IT OVER

THE UNITED STATES FOREST SERVICE RETIREES OF THE IN-
termountain Region, headquartered in Ogden, Utah, fifteen
years ago formed the "Old Timers Club." Purposes were to
maintain contact between retirees and keep a record of ad-
dresses. They distributed periodically the "Old Timers
News." Members were urged to write of their experiences
so the club could preserve the material as historical records.

Having served for ten years in the Intermountain Region,
beginning when the Forest Service was only seven years old,
I recorded incidents that seemed of interest but I hadn't made
them available.

I read *Gold and Cattle Country,* by Herman Oliver, col-
laborating with E. R. Jackman, and I later read *The Oregon
Desert,* by E. R. Jackman and Reub Long. I could hardly
lay the book down. Near the end, I read almost all night.
In the morning I got in touch with Russell Jackman, whom I
knew somewhat. If he could cooperate with a cattleman-
banker and with a horseman-buckaroo-rancher-philosopher,
why couldn't he help a fire-fighting forester with forty years
of "rangerin."

This manuscript is the result. It is not a biography, or a
complete history of the Forest Service. It is a glimpse of the
United States Forest Service over a fifty-year period as lived
and observed by one who worked on the forests and ranges
of Minnesota, Idaho, Oregon, Washington, Montana, Wyoming, Utah, Nevada, and Arizona. I have traveled on gov-

ernment business by shanks' mare, saddle horse, muleback, buckboard, canoe, whitetop, lumber wagon, bobsled, horse stage, Model T, Jeep, helicopter, and trimotor. I have rubbed shoulders alike with lumberjack and lumber company president, sheepherder, cowboy and stockman, homesteader and bank president, prospector and mining executive, smokechaser and Chief Forester.

No first-name basis here, but I served under seven Presidents of the United States. The Republicans had it for numbers, but the Democrats had more years in office. They didn't seem much different to a forester way out in the Rocky Mountains. I worked under the supervision of seven Regional Foresters.

This seal has been adopted as a symbol of multiple use. It is based on an ancient symbol for wood. Six elements make up the symbolic "Multiple Use Tree," the central theme of the design. The branches stand for the five major resources of the national forests—WATER, WOOD, FORAGE, RECREATION, and WILDLIFE. The trunk stands for the nation and its people who benefit from the use of these resources. Line continuity symbolizes Multiple Use Management as practiced on all national forests.

Yet, after all, this is only a more genteel way of saying what the pioneers on the Old Oregon Trail, the early-day forest rangers and other mountain travelers, looked for daily— "wood, water and hoss feed." Recreation was represented usually by rest following a strenuous day. Wildlife sometimes contributed a skillet of wild meat. Multiple use is therefore not new. It is merely more stylish in a new suit of clothes.

Early-day problems, experiences and observations involving these multiple-use resources make up this book. In the early days utilization was limited and custodial work was major. Forest fires, once referred to as "The Forests' Prime Evil," are one of the most dramatic and basic. New activities developed through use and management.

Forest homesteads were an important and interesting human factor in the settlement of the West. Land acquisition and exchange form the basis of one chapter. This activity simplified the management of the forests.

An inseparable part of the management and use of the forests is the personnel. Training young foresters is described, going back almost to the turn of the century. During the depression days of the thirties the CCC program was born, and intimate contact with this new development is recounted. Having worked where many wood ticks were infected with Rocky Mountain spotted fever and having sat by the bedside of a stricken fellow worker during his dying hours, this subject seems important.

"Survival: Keep Cool and Live" is offered as a service to

the increasing thousands of visitors who are going for the first time into strange mountain country. This chapter could save human lives.

"Burros, Buckboard, Bedrolls, and Beans" tells of observations and experiences, unrelated to anything else, which throw light on the doings of government foresters.

Two chapters, "Trees and Towns" and "Trees and Humans," were written by E. R. Jackman. In them he shows how forests benefit the communities in the region. He shows the effect of woods upon individuals and why persons of all kinds like to go into Western timbered mountain country.

We are indebted to Mrs. Alice Hutchens, of Kalispell, Montana, sister of Mr. Jackman, for the final chapter, "My Friends, the Trees." She lived for several years on a forest homestead in the former Blackfeet Forest in Montana and is a true lover of the trees and flowers and wild things growing in the woods. She paints beautiful word pictures of them. In addition she gives her own homesteading experience in the chapter "A Bet with Uncle Sam." She wrote the section "Lumberjacks I've Known" in "Timber Oup the Hill!"

E. R. Jackman has masterminded all of the chapters. He proposed the general makeup. He edited every page, substituting a four-letter word for a longer one now and then, smoothing out my sentences, weaning me from my bureaucratic style, and repeating that "Knowledge need not be dull." He always had the reader in mind, so clarity and readability were his watchwords. He tells of the Idaho fan whose letter in full was "your ritin makes rite good readin."

Both of us were born in Minnesota, both have worked our mature lives serving the public in the West. We have many things in common, so we made a congenial team.

I offer the book with the hope that it will appeal to persons who have an interest in forests, whether personal, commercial, or official. There are chapters for each. The chapters, taken together, tell much of the story of the United States Forest Service.

E. R. JACKMAN

An agronomist, well known everywhere in Oregon and even beyond its borders, he is now retired after a lifetime with the Oregon Extension Service. He is noted for his work on forage crops. His writing is easy to read. Even in scientific matters he has an everyday touch and his own kind of fetching humor.

I suppose that every man worth his salt likes to think that his life has had meaning and worth. It has been a full life; I think it has been useful. There has been no personal unemployment problem, for a day was never long enough, and the years raced by. I have never understood persons who were bored or who hated their work. Above all, my work was interesting to me and enjoyable.

I hope some of those things show through in the various chapters. If the book adds to the reader's knowledge of the forests, that is something worthwhile.

CHAPTER TWO

A BET WITH UNCLE SAM

By Act of June 11, 1906, Congress authorized homesteading of national forest land, provided the tracts were chiefly valuable for agriculture. Some of the forest boundaries had been drawn with a broad brush, while flats and stringers of good soil were sometimes found along the watercourses. Some level to rolling benchlands had possibilities for raising crops. The large percentage of land was too steep or too rocky or at too high an elevation for farming and was principally useful for production of timber, for range use, for watershed protection or recreation.

The law provided that upon request of a would-be homesteader for a certain tract, the area must be examined and, if found suitable, would be "listed" or opened for entry. Thereafter it could be filed upon, the applicant having a sixty-day preference, after which it was available to anyone. The requirements of cultivation, improvement, and residence having been met, the homesteader could "prove up" on his claim. In time his patent would be forthcoming. A patent corresponds to a deed but it comes from the United States government and is signed by the President of the United States.

An advantage of a "June 11th Claim," as they were called, was that they could be located on land not yet sectionized or surveyed by the Land Office of the Department of the Interior. They could be located by metes and bounds so long as no part of the 160 acres extended outside of a square mile.

With the use of a compass mounted on a tripod for determining the courses, and a surveyor's chain for measuring distances, an irregular tract could be laid out to include the desirable farmland and to exclude rough or unwanted areas. Survey and examination were done by a forest officer but with the applicant present if possible. Each of the several corners was marked with a stone, a post or a scribe mark on a blaze on a tree. This was known as the listing survey.

After these irregular-shaped tracts were filed on, it was necessary to make a precise transit survey and establish permanent corners. Only a licensed surveyor could do this. It was called the "entry" survey. This farm boy from Mee'-na so' tah' broke in on this job.

On July 9, 1913, I rode the Oakley branch train from Minidoka to Oakley to report for work on the Minidoka Forest in Idaho. On the same car I spotted a fellow I thought might be a Forest Service man. He wore one of those stiff-brimmed Stetsons and leather puttees and had a transit case.

I found lodging, then learned where the Supervisor's office was and went there to meet Supervisor William McCoy. Talking with him was the man I had spotted on the train. He was Ben Rice, years later to become Regional Forester. He came to make the entry surveys of some forest homesteads. The next day I went with him as chainman and assistant. That evening before dark he set up the transit and got ready to get a shot on Polaris at its western elongation. This was scheduled in the Engineering Handbook to take place at 11:00 P.M. so we went with flashlights about 10:00 P.M. to be on time. At 11:00 P.M., the star was still going west. Checks through the telescope every minute or two found it still moving west. And although it was summer the mountain air chilled us to the marrow. Slightly after midnight the old star stopped and then began its eastward movement. A peg was set and tack placed to mark the line of sight and establish the true meridian. Only then could the transit be put away and 1:00 A.M. saw us in our camp beds. So ended my first day under Civil Service.

On surveyed land (sectionized) the prospective farmstead would be described by any legal subdivisions as a 40, a 10-acre tract, or even by 2½-acre squares so long as all the subdivisions were contiguous. This made it possible to allow the applicant the largest percent of usable ground in patchy areas. Sometimes the boundaries shown on a map looked like a stairway to a high second story. The number of corner posts needed in fencing such a claim was staggering.

In the 1912-15 era a land scramble occurred in southern Idaho. June 11th Claims were in demand not only from local people, but by voyageurs from Oklahoma, Arkansas, and other Midwestern states. The policy was liberal. The stouthearted candidate hardy enough to believe he could make a living on the claim, got his chance. The area was withheld only if it was necessary for an administrative site, or for a sawmill setting, or if it included a water hole needed for stock, or for a roundup or holding ground for cattle.

For several years homestead work was one of my assignments. There was only one appeal. In this case the applica-

Photo by Charles Simpson, Baker, Ore.
A BACHELOR'S CASTLE
No clothesline here. The aspen, deformed by winters' snows, show that this upland country was probably at an elevation too high for farming anyhow.

tion was for 160 acres on a big plateau at an elevation of 5,600 feet and right smack in the middle of a good cattle allotment. A short growing season and the probability of frost damage made farming questionable. The recommendation against opening to homesteading was approved by the Forest Supervisor and Regional Forester but was appealed by the applicant to the Secretary of Agriculture. In due time a Mr. Shearman, direct from Washington, D.C., arrived. On July 10 he looked over the ground in company with the Forest Supervisor and his assistant, a couple of stockmen, the applicant, and four witnesses for him. It was one of those blustery days and while on the site a brisk snow squall hit us. Everyone was wearing his chicken-skin clothing and one chap commented adversely about the July weather. One of the applicant's witnesses, thinking to be helpful to his side, said, "Think nothing of it—it's apt to snow up here any month of the year." The claim was disallowed.

Thinking of homesteads always brings to mind Arthur Summer. He had a June 11th Claim in the Beaverdam country southwest of Oakley, Idaho, but by some miscalculation he had only 130 acres and wished an addition of 30 acres. So it was surveyed out in the forenoon and we were invited to stay for dinner. He and his family were living on the claim in a tent 14' by 16'. His wife had gone to town and Summer cooked dinner, including some extra-fine home-grown fried chicken. It was in midsummer and the dirt floor was quite a dust bed. A big, yellow, long-haired, long-tailed dog would lie in the dust, then walk close to the table wagging his tail, and the dust would fly. Some half-grown chickens had the run of the tent and when disturbed they really made a dust cloud. A tame magpie would light on the shoulder of one of the youngsters and receive a chicken bone. Then he would fly to the water bucket sitting on a barrel head, light on one side of the bucket, then hop across to the other. On top of the warming oven was a quart fruit jar of coyote scent. This is made from glands and other parts of the female coyote. It was working and the bloody bubbles would zigzag from near the bottom up along the side of the jar to the top. I'm still fond of country-fried chicken.

Many of the homesteads around the foothills were to be dry farms and many had no domestic water. This meant hauling water in barrels with team and wagon. A big spring on the north end of Black Pine Division alone provided water for a dozen homestead families. Baths were few and far between. I knew one family that hauled water for seven years. Finally they got around to digging a well and hit water at fifteen feet.

Just east of Strevell, Idaho, close to the Utah line, a young fellow applied for a quarter section of foothill land inside the forest boundary. While we were looking it over the applicant offered me one hundred dollars if I would get it listed for him. I advised him that if we found the tract suitable it would not cost him anything, but if the findings were adverse it was our job to report it that way. That was my

Photo by Charles Simpson, Baker, Ore.
U.S. RURAL POST OFFICE—1913 MODEL—SOUTHERN IDAHO
Isolated post offices were usually right in the homes of homesteaders. You could ride in, get your mail and a welcome cup of coffee all at the same time.

only bribe offer in a long career. Sure, I've been given a leg of lamb or a piece of mutton a few times, and once a nice box of apples arrived from a user in the apple country. Once I found a quart of seven-year-old whiskey on my desk. Actually that did me little good and I never knew who the donor was. Such things raise a nice question. Where do you draw the line? Really, they were purely acts of friendship. Most every forest officer has delivered mail, relayed messages, reported stray sheep or an injured cow to the owner, or helped a logger. No reward is expected. But who wouldn't want to show his appreciation?

When applications for homesteads tapered off, a classification of the whole forest was made. This divided the forest lands into two classes: (1) known nonagricultural and (2) possible agricultural value—to be determined later. Some of this work was done in connection with a boundary survey and boundary marking job. With a three-man crew and pack outfit we were doing the three jobs in one, viz: survey, signing, and land examination. We'd been out on the job some time and the tobacco ran out. One man missed it badly. We noticed that he searched his pockets for tobacco crumbs, even turning all his pockets inside out and getting enough from the seams to roll a skimpy cigarette. The next day he quietly pulled out afoot for the nearest supply point and didn't return. Jobs were not too plentiful either.

On this same project we were camped along the boundary a couple of miles from a homestead. When we returned to camp one night after a day on the line, I threw back the flap of the cook tent and "Woof, Woof," a big old sow and a litter of good-sized weaner pigs charged out, almost upsetting me. And was it a mess! Our flour was scattered and eaten, our bacon devoured. There just wasn't anything left fit to eat. Our dishes and pots and pans were strewn about and messed up. I had been sleeping in the cook tent and my bed was practically ruined. A .22 rifle had stood in the corner but it was down and mixed with filth. I've seen camps molested by porcupines and bears but never anything to compare with this.

The areas left for intensive classification were examined jointly by a forest officer and a representative of the Bureau of Soils (forerunner of the Soil Conservation Service). A chap named Gunderson was detailed to our job from the Bureau. He was a fine guy and knew his soils and I learned a lot from him. Yet, like most of us, he had his faults. I had a fine team of horses, half brothers, bright bay with blaze faces. They were broken to harness, but also were good saddle animals. I drove them with a white-top buggy. Gundy and I had agreed to travel light as our space was limited and we had to haul two saddles and some oats for the horses. Our ideas for traveling light didn't coincide. He had enough clothes to last for a year, winter and summer, rain and shine. Included were three pairs of boots and other things in proportion. What I couldn't stow away I cargoed up and lashed on and we took off. On occasion we would have to leave the buggy, saddle the horses, and ride cross country. Some roads had to be covered horseback and any road mile seems longer than a cross-country mile. I'd learned from my cowboy friends that an easy, swinging trot is the best mile eater for both man and horse. After the first half mile Gundy balked and pulled his horse to a walk. I jogged along and when I was half a mile ahead of him he went into a lope and passed me, then after a time slowed to a walk. We continued this zig-zag way to camp. I've always been a little ashamed of my part in that exercise. But prior to that he complained bitterly that his horse was the roughest horse he had ever ridden. I'd exchanged horses, giving him the one I was riding. But he was still unhappy.

Some of the parcels opened for homesteading were never filed on. In other cases the homesteader found it just too tough and gave up. Lack of capital was the biggest problem. Schools were miles away and there were no busses in those days. Living quarters had to be built, fences were needed, sagebrush and juniper had to be removed, seed must be purchased. Rabbits and squirrels were numerous and damaging. Equipment was necessary. Feed had to be produced before livestock could be kept. Stock cost money. Progress was pitifully slow. Homesteaders could move off the claim for five months each year, and many did. But they could earn only enough to pay expenses, with little left to develop the claim. Many became discouraged and abandoned the homesteads. Such tracts remained open for new filings. If no one else filed after a reasonable time, the listing was recalled and the land returned to national forest status. Of those that went to patent, a small part became headquarters for successful small stock ranches. These usually were on a creek, with some good bottom soil or land that could be irrigated. A larger percentage were sold to valley stockmen who used the homestead only for pasture. Some went tax delinquent and became the property of the counties.

The Forest Homestead program in southern Idaho and elsewhere proved not to be a very pretty picture. Hardship on the part of families was endured. Hard work and some cash went down the drain. It was a phase in the growing up of the West. Part was the result of the too common human desire to get something for nothing. My own grandmother lamented to her dying day (and she lived to be ninety-six) that she had not been able to use her homestead right.

One homestead that succeeded was taken by William Gunnell on the east side of the Raft River Valley at the forest boundary. A small stream provided water to irrigate hay land along the creek. The place, known as the Gunnell ranch, became the headquarters of a small going cow ranch. It supported a year-long family home and three sons grew to maturity on it. Bill told this story:

The ranch was seven miles from the dirt-roofed post office at Bridge, Idaho. That was the nearest telephone. The Gunnells decided to build their own line to Bridge in order to save a lot of riding. The three nearly grown boys started work. Their dad said they got one mile of holes dug and ran out of material.

In Bear Lake County, Idaho, an old-country cheese operation was carried on in a high mountain meadow within the Caribou National Forest. It was on deeded ground owned by three Swiss families named Kunz. The three brothers had farms in the valley at Bern, located a short distance westerly from Montpelier. Every summer they would move their cows and young calves to the big meadow where each had a cabin and cheese plant. At each of these summer homes were two round pole corrals, each large enough to hold sixty-five to seventy-five cows with their calves. The corrals adjoined, with a connecting gate. Dad, Mother, and five or six youngsters made up the milking crew at each place. At milking time the cows would be in one corral and the calves in the other. One at a time each milker would let a calf through the gate, then with milk bucket in hand and armed with a stout stick a foot long, he would follow the calf to its mother. After the milk was well started the calf would receive some sharp taps on the chin or jaw. This was his cue to wait for the second table. The milker would take about half of the supply while the calf waited impatiently. The milk was carried to a big vat and the milker returned to the gate for another calf. The cows went out on the meadow during the day and were brought in at evening for another milking. Mother and calf were separated again overnight. Morning milk was added to the milk already in the vat and cheese making started. An empty vat was ready for the next evening's milk. The big cartwheels of Swiss cheese were taken to market periodically with team and wagon.

A modern version of the desire to take homesteads developed in 1948 in the Whitman National Forest, Oregon. Applications were received for fourteen quarter sections (2,240 acres) forming a single block. It was located on Spring Creek, Union County, about halfway between La-Grande and Meacham. Two thirds of the total area supported a good stand of second-growth yellow pine, the balance was good grassland with scattering browse. Soil varied from good forest loam to rocky ridges and benches. A stream crossed the block and there were three or four springs, but some of the individual units were completely dry. It made up part of a fenced cattle allotment.

We arranged for the Soil Conservation Service to make a soil survey and map. The Forest Service typed the timber and other kinds of cover, estimated the volume of timber of commercial size and the stand of young trees per acre. The county commissioners looked over the area and, considering the need for roads and school facilities and value for farming purposes, recommended against opening for homesteading. Pictures were taken of each tract applied for. A report of all the findings was sent to the Regional Forester recommending against opening. He sustained our finding.

On every forest there were a few fellows who were a little queer. It didn't mean a thing to them when the President pointed out that they were ill-fed, ill-clothed and ill-housed. They would rather be out there and be independent than to have a good job and have someone telling them where to pile their hay. Some trapped, some prospected, some, as a cowboy expressed it, "just resisted." But they got something out of living in the forest and they got happiness of a sort from their homesteads.

These men grew old, and from society's standpoint it would have been better had the homestead never been taken, for often a succession of other owners came, all distinguished as failures.

With improved roads nearly everywhere now it is hard to understand what some of these homesteaders went through. Here is a letter from Alice Hutchens, of Kalispell, Montana, who lived on a forest homestead about fifty miles north of Kalispell on the Blackfeet Forest.

"Homesteading forty miles back in the woods had its hardships, but it had its compensations, too. For eight months

Photo by Pendleton Grain Growers Co-operative

THE LAZINKA RANCH HOME, UKIAH, OREGON

Yellow pine trees, a spacious grassy yard, and a clear, winding creek provide a gracious and charming background. A Polish immigrant homesteaded here, surrounded by forest land, and succeeded.

of the year, the road, that was barely passable at any time, was apt to be completely closed to wheel traffic. Sometimes, for weeks, in deep snow, no horse could negotiate it, and often snow conditions made even snowshoeing impossible.

"The year young Jack was born was such a one. Star Meadows had over ten feet of snow, measured as it fell, and over six feet of packed snow by the time spring came. I was 'out' most of that winter but when Jack was five weeks old, spring had come and we knew if we didn't get home before the rivers came up, we might not get there until July.

"We had 48 miles to go. The first day took us, on wheels, to the edge of the timber. The second day we moved everything onto a bobsled. We had reached the snow country. Snow was six feet deep in the timber. A few people had made sled trips out from time to time, and the snow was packed in two sled tracks about four feet above ground, with two feet of crusted snow on each side of the tracks. As long as the tracks were frozen, going was hard but possible.

"On the second day, about five miles into the woods, a warm rain started and the bottom began to go out of the tracks. The sled slipped off, first on one side, then the other, and the horses began to break through. After two miles of this and most of a day of digging out the sled and resting the horses, we reached the homestead cabin of a young bachelor we knew. He was glad enough to have us stay the night but was mortally afraid of the baby. We repayed him with jars of home-canned jellies and such and with some of the potatoes and vegetables and eggs we had brought along.

"The third morning we started long before daylight, while the road was frozen. There was a four-mile stretch of up-hill road. We couldn't make it quite to the top. The road was thawing again, and the horses were getting very tired. We unloaded all my boxes of home-canned things, a cedar chest, a chest of drawers and most of a three-month supply of groceries we had and piled them under some thick spruce trees that gave quite good protection from the weather. We left the sled box and back bobs there too. We built a platform of poles on the front bobs and went on, with a few groceries, some bedding, and the baby's clothes. We walked so that the horses weren't hampered by our weight on the sled. I tried riding one of the saddle horses we had along, and carrying the baby, but what with the horse slipping off the track or falling through it, then plunging and struggling to get back on firm going, I was better off walking. We got to the top of the hill, stopped to rest the horses and feed them oats, then went on. There was still a five-mile journey before we would reach the Talley Lake Ranger Station where we could spend the night. The downhill road was on the shady side of the mountain and wasn't bad. We had to ford the river, and that meant unpacking the things from the sled again and carrying them across a foot bridge above the ford. The water was over the top of the platform we had on the sled. We reached the ranger station about dark. Homesteaders were all kept advised as to where to find the key, since emergencies such as our were common.

"Before daylight we started on the last eight miles. The road grew worse. The horses gave out. The going was impossible for them. So we unharnessed them, left a collar on one of them so we could tie a bundle of baby clothes to it, let them rest an hour, and started up the long, long hill that was Lookout Mountain. We tried getting the horses to go ahead. They simply wouldn't face it. So I went ahead, coaxed, petted, talked them into getting up again the first few times they fell—and finally they followed me. By resting every few rods, we all got to the top, Jack coming along behind carrying the baby on a pillow.

"From there on, we were in dense timber where the sun and wind hadn't made much headway and the road held us up. We finally got to a barn where we had hay. It was getting dark by that time. Jack got the barn door open someway and we just turned the horses in and left them.

"The Sinclairs lived about two miles beyond, and she had been expecting us for four days, and had the house breath-

lessly hot because she felt that a new baby was a fragile creature. I guess ours wasn't.

"All next day Jack spent making a trail to our own cabin.
No one had been up that way all winter except for a neighbor on beyond our place who had moved out early in the winter. They had shot the lock off the door and had lived in our
house while they packed and moved their things. They must
have been there two or three weeks. They'd used up every
bit of our food, slept in our beds, used up our woodpile, and
there wasn't a clean dish in our house.

"So *that* was our homecoming! Out went everything
they'd used except the dishes. We spent our first two hours
cleaning the place, and Jack had to go back to Sinclairs and
borrow food. We had clean things left in bureau drawers.
Jack went on snowshoes next day and back-packed what
food and clothing we had left at the foot of the mountain
and a great deal of sudsing and boiling went on out-of-doors
until everything was clean again. This was the unglamorous
side of homesteading.

"But there were compensations—clean air, clean streams,
beauty everywhere we looked, cold springs of soft water,
birds, animals, flowers, ferns, grasses, berries for the picking,
wonderfully productive soil for gardening, good fishing and
hunting and time to study and enjoy it all.

"I was alone a lot, but I had a small library of very good
books, among them a large, thick book called Comstock's
Natural History and the entomology book my brother had
used in college. We got the Kalispell Bee and the Country
Gentleman. Sometimes there would be two months between
mail arrival, but we and the Sinclairs always collected each
others mail.

"I liked to sew, and made all of my own and the baby's
clothing by hand for the three years I lived there. I'd had
rather nice clothes of good materials, as I'd been a teacher.
They were made over and made over again.

"We cut hay on a large meadow two miles from home.
All three of us went every day. I took the baby along down

to the river in his buggy and could always catch fish enough for dinner, cooked over a campfire. Then, in the late afternoon I'd go again and catch enough for supper and breakfast.

"Once I lost the baby! He was sleeping, with Old Bill beside the buggy. I moved about ten feet along the riverbank to a nice fishing hole, got all the fish we needed, and came out the wrong deer trail. In ten seconds I was hopelessly lost. The more I tried to find the baby, the more confused I grew. I couldn't even find the *river*. I realized that I was being senseless about things and stood still until I could hear the faint noise of the mower out on the meadow. I kept stopping and listening for it, and finally got out of those dense willows into the open. Then it was easy to find the trail I always used, and go back to the riverbank. There was the baby, still asleep, with Old Bill Airedale sitting by him, resting his big old beautiful head across the foot of the buggy. Good Old Bill! Bless the faithful heart of him! If *I* wasn't going to take proper care of my child, *he* would look after him.

"On hot July afternoons I often took the baby along a trail at the foot of the hill to the edge of a cool green thicket, and while the baby slept or amused himself watching the play of sunshine and shadow on the leaves, I picked wild strawberries or little alpine huckleberries along the hillside. They were sweet and flavorful but so small that it took most of the afternoon to get enough for a meal. It was a quiet, peaceful place and I felt ever so safe there.

"In the fall Jack set several coyote traps in that thicket and a bear got into the whole lot of sets. He had a trap on every foot. Naturally he was pretty peeved about it, and reached up and bit chunks of bark off the trees higher up than Jack could reach. Those thickets were so thick you couldn't see six feet into them, so Jack sent Old Bill in to find out which way the bear had gone. There was considerable scrambling around among the bushes, then out came Old Bill, walleyed and staggering. The bear was in there, all right. He'd walloped Old Bill over the head with a trap. Later Jack found all the traps—the bear had tangled the trap toggles in the bushes and pulled his feet free. I'm glad he did. Probably that huge old bear had spent many a long idle afternoon watching us. Many signs said that he must have spent the whole summer there.

"Another time I walked up on a golden eagle eating a fawn. He ruffled up his feathers, raised his great wings and glared horribly. I got out of there, trying hard to look as if I hadn't even been there.

"Only once did I ever have any experience with a lost person. The baby and I were home alone. Jack was upriver with a hunting party. There was no snow, but in the night we had several hours of rain, then it turned very cold. At three o'clock in the morning someone knocked on the door. It was a young fellow, 17 I guessed. He had been lost most of the day before and all night. He was wet through and his outer clothing was icy. He was utterly lost—had no idea where they'd started into the woods. I built up a hot fire, made coffee and cooked breakfast as he hadn't eaten since early the day before. Meanwhile he kept eyeing Jack's Luger that just happened to be on the table. He asked if I could shoot it. I said I'd killed a few hawks with it. He shuddered. At the first sign of daylight, he wanted to leave. He couldn't keep his eyes off that Luger. So I headed him down the wagon road and told him to stay on it until he got to Sinclairs. He did. They took him to the Ranger Station and his party found him there.

"I don't know just how Jack found his way in the woods. It seemed to me that he just looked around at the lay of the land, kept in mind a map of the hills and gullies, picked out cliffs, bare spots on hills, old burns, and any other landmarks that showed up for a long way, and carried it all in his mind. It must have been as plain to him as a printed map would be to me. He was never lost, even in fog. He was in the hills a lot and noticed all sorts of small things that marked that special trail. Maybe he had a built-in sense of direction. Son

Jack did, even before he could talk. No matter where we were, he could always point out the direction toward home. He has the same ability to make maps in his mind, and map reading came natural to him. Instead of getting his basic training in the Army, he spent most of his time teaching map reading.

"Neighbors were few and far between. Nearest us were the Sinclairs. They were Scottish people, and good neighbors. Mrs. Sinclair was the only woman in the Meadows excepting myself. Mr. and Mrs. Sinclair spent winters out, but Uncle Hector and Cousin Urquhart stayed to take care of the cattle. The Sinclairs weren't homesteaders. They bought a large tract of meadowland.

"Albert Jones and Charles Oehteker were two bachelors who had one fork of the river to themselves. There was no

Courtesy Malheur National Forest
MOUNTAIN SNOW OFTEN ADDED TO THE BURDENS OF THE
NATIONAL FOREST HOMESTEADER

road up their way. They lived quite a distance apart and neither had a horse—not even a dog. They were both fine-looking men, well groomed and clean at all times and were meticulous housekeepers. They were in the woods for the same reason—they were living under a self-imposed exile. Jones was a graduate of a fine Virginia University. Oehteker was graduated from a well-known university in Austria. Jones had disgraced himself by being let out of a government job for gambling. Oehteker left Austria to escape military service. For months at a time they saw no one but each other, and yet each rather despised the other because of his reason for being there. They trapped for what money they needed. They both had flourishing gardens, and of course the woods and streams provided meat. One spring when travel was impossible, they lived for five weeks on fish and rhubarb. They were none the worse. We had a small flock of hens and they used to bring us rhubarb—beautiful, huge tender stalks too long for a gunny bag to hold and we gave them eggs. They said it had been years since they'd had hens' eggs, and that it was hard to find good fresh wild ducks' eggs. Jones held up well under the isolation, but Oehteker became queer. One day they were walking along a trail together when Oehteker suddenly grabbed Jones' rifle, bent the barrel around a tree, and went running off up the trail. Jones didn't linger around there. He set out at once for civilization and went to work in the hayfields. No one saw Oehteker for two months—his garden grew up to weeds and his much-loved pansy bed bloomed itself out. Then he reappeared, looking well but shaggy and declared, 'A man doesn't have to eat. We only eat because we enjoy it.' That's all anyone ever found out. Jones, when he became old and ill, shut himself into his house and deliberately drank himself to death. Oehteker is still alive—very old with next to no memory, very afraid of strangers, and almost helpless. His place went for taxes years ago, but he doesn't know it. The Sprouls, who own all of Star Meadows now, look after him. Anyone else would have had him taken to the County

Home years ago, but the Sprouls know that would break his heart, so they have undertaken the care of him although it is several miles up to his place.

"You'd like the Sprouls. He comes from an old-time Wyoming cattle ranching family. Her people from away back had a sort of inn and trading depot for wagon trains, and later a stopping place for stagecoaches in southwestern Wyoming. They are along in years, but such happy people! The boys run the ranch. They have Hereford cattle and Hampshire sheep. They saw the write-up of The Oregon Desert in the Farm Journal, sent for the book, and almost know it by heart.

"Then there was Mr. Huber, quite a neighborly sort. He spoke English, but used so many German words along with it that I had to get used to him before I could visit with him. He used to happen along every few days during the summer, his pockets bulging with mushrooms. He hadn't been long away from Germany—later on, in World War I, he had a son in each army—German and American. He took no sides at all. He had a huge young Clydesdale stallion that he'd raised from a colt and made a great pet of. He called it by a German name meaning 'handsome' and he had taught it to kiss him. One day he was leaning over the well-curbing getting up water. He felt Handsome nudging him and said, 'Oh, did you want to kiss me?' The horse leaned over the well curb too, and kissed him with such affection that he broke Mr. Huber's jaw in two places. He never blamed the horse.

"Freeland and Diller lived up still another creek. They worked out some at ranches and logging camps. No one liked them—no one trusted them—no one dared to call on them because they were known to have set deer traps in the trails and claimed to have set guns, too. (Deer traps are awful things. They are rather like coyote traps but have sharp steel spikes in the jaws. Jack found one once, after they'd gone from the Meadows.)

"When about a half mile of Mr. Huber's woven wire

fence turned up missing, they were finally investigated. About every article anyone in Star Meadows had ever lost was found there. They'd also been moonshining. They were both in the penitentiary the last I knew. Freeland had been hiding out for years for some crime he had committed long ago in South Dakota.

"Then there was Old Man Swanson, a gloomy old soul. A visit from him was about like sipping vinegar, but he did carve a nice little farm out of the woods, and his house and outbuildings were beautifully planned and built. He had been a boat builder in Sweden. His whole place was neat and attractive. He was very old when we knew him. He was neat and clean right up to the end. They found him dead leaning over his washtubs.

"These were the neighbors—*all* the neighbors."

CHAPTER THREE

TRAINING YOUNG FORESTERS

Civil Service

ALL PERMANENT EMPLOYEES OF THE UNITED STATES FORest Service are under Civil Service appointment. To qualify they must pass an examination, changed frequently. Each type of work has a different examination.

The early-day ranger examination was a practical, out-in-the-woods test of a man's ability to hold his own. He had to bridle and saddle a horse, first fitting each, then mount and move the horse around a bit. The examiner could instantly distinguish the greenhorn from the horseman.

A ranger had a large territory to cover with no motels at convenient, one-day intervals. He had to take with him his roof, bed, kitchen, groceries, and probably some oats for the horses. The candidate would approach a formidable array of supplies with instructions to load it onto a gentle mule or packhorse. This heap included blankets, an axe and long-handled shovel, a couple of frying pans, a coffeepot, water bucket, odds and ends of dishes, a couple of small wooden boxes with two weeks' supplies of groceries, a coal-oil lantern, half a sack of oats, a sawbuck packsaddle, rope, and a canvas fly. Usually, but not necessarily, a pair of panniers or pack bags were provided. The trick was to assemble this, put it on the mule, tie it on, then lead the mule around for a few turns. Either the candidate knew the answer or he didn't. Packing a mule resembles driving sixteen mules with a jerk line. You can do it or you can't.

There are several ways to put on the lash rope. Among them are the double diamond, the single diamond, the squaw hitch, the sheepherder hitch, and numerous variations. The single diamond is used the most. In the northern Rocky Mountain country an entirely different pack was regularly in use. Instead of the sawbuck saddle, a Decker saddle was employed. The trees were shaped to better fit an animal's back than those of the sawbuck. They were connected by two shaped iron rods, one fore and one aft. A bow or arch provided lots of room to clear the animal's back and gave a place for tying the sling ropes. A filled canvas pad fitted down on each side on top of the tree with the arch of the tree irons projecting through two holes in the pad. Saddle blankets and pad were used next to the mule's back. Two canvas manties 6′ x 8′ were used. The load for each mule was divided in half, placed compactly on the manties, and wrapped and tied with a cargo rope. If possible, these packages were made longer than they were wide, about the shape of a large egg crate. Two sling ropes were tied permanently to the front arch of the saddle. When one cargo was laid up on the animal's side, the sling rope went across the pack through the rear ring or arch, down under the end of the pack, up and tied to the rope that crossed the upper part of pack. No top pack or extra cinch or lash rope was used. To balance the load, which rode high on the mule's sides rather than on his back, it was a simple matter to raise one pack or lower the other. This Decker saddle* and method of packing is simpler and easier to learn than other methods.

The third test was with an axe. A tree about a foot through was selected. A stake was driven into the ground fifty feet from the tree. A double-bitted axe was handed the candidate. An axeman is known by his chips. If the stump indicated a beaver had chewed off the tree or if it toppled in the opposite direction, he didn't need a second guess as to his grade.

---

* See E. R. Jackman and R. A. Long, *The Oregon Desert* (Caldwell, Idaho: The Caxton Printers, Ltd., 1964), pp. 265-70.

Pistol shooting was the final test. The target was a gallon can on a stump at fifty yards and another at one hundred yards. The candidate, standing, used a pistol (his own if preferred). He could shoot from the hip if he wished. Then he was given three tries at a tomato can tossed into the air.

These tests did more than show the applicant's prowess in the woods. The examiner had a good opportunity to size up the candidate: his whole attitude; his willingness to try; his ego or lack of it; how he accepted failure; his perseverance; his thoroughness. Some men were selected even though they did get the packsaddle on hind end foremost. It was important to be in harmony with the mule.

Before long the woods tests gave way to the ranger's written examination. This was a short, one-day test. It covered

Courtesy U.S. Forest Service
PACKING FOOD AND EQUIPMENT ON A MULE
This was the principal mode of transportation in 1910, and every forester still has to use it on occasion, even in 1967. Shown is a Decker saddle, with which the load is made up on the ground, a separate pack on each side.

a wider list of subjects designed to determine the candidate's practical knowledge of timber work, grazing business, fire, surveying, and other subjects. A part of one test read, "Give in detail the difference between handling sheep and cattle on the range." One examinee answered, "Cattle on horseback— sheep on foot." Sime Kinney went to town from camp to take the exam and in the timber and sawmill section he encountered this question, "What is a steam nigger?" He put down, "That's that greasy old cook they have up at the planting camp."

Questions now and then might include definitions for: "undercut," "pulaski," "base line," "blaze," "peavey," "closed area," "cruise," "deadfall," "schoolmarm," "alforjas," "bannock."

In Nevada and southeastern Oregon years back occurred a serious outbreak of rabies. Public and private organizations were alerted to locate tame and wild animals with symptoms. Now I can't vouch for the truth of this story but one person said an examination was given to an Irish boy and in it he found, "What are rabies and what do you do about them?" His forthright answer was, "Rabies is Jewish priests and yez can't do a damn thing about thim."

In many ways the written test was better than the practical test but it had its disadvantages. The selecting officer had no opportunity to size up the man before appointment. Too, a person with reasonable intelligence and some study could pass with a fair grade.

A list of three eligibles would be furnished a Forest Supervisor, and he was required to select the top man unless he could justify passing over him and picking No. 2. If No. 1 was known to be a bad actor or a crooked bootlegger he could be bypassed. The same applied if No. 2 had a smelly record. In those days there were no personnel officers to expert the selection. Supervisor Grandjean of the Boise Forest had a clerical vacancy. He was supplied a register showing three girls in 1-2-3 order of their ratings. He picked out No. 3 and got her on the job only to be questioned by higher authority as to why he had not selected No. 1. His answer was, "Well, I didn't know any of the three, but I thought from the accompanying pictures Miss 3 would fit into our organization better than Mrs. 1."

On the Caribou Forest a vacancy existed for a district ranger. We were provided with an eligible list of three and selected No. 1. He was a grade-school teacher and had a high rating. But he hadn't had a particle of timber or live-stock experience. His first job was counting sheep. A main stock driveway came through his district and many of our permitted sheep entered there as well as a large number of bands headed for a neighboring forest. One hundred and twenty-five thousand head were counted there. I went out with him and got him started at the counting chutes. In a couple of days I phoned and asked how things were going. He said, "Well, I'm learning damn fast to count sheep in the daytime—but you didn't tell me about counting sheep in my sleep."

Oddly enough quite a few men from other vocations were interested in Forest Service employment, took the ranger examination, passed, and became rangers. Supervisor Mc-Cain, on the Teton Forest at Jackson, Wyoming, used to be proud of the assortment of men he had. Felix Buckenroth had been a captain in the regular army. Another man had been a practicing veterinarian. A third, Dibble, had been a civil engineer (C.E. degree) with the Bureau of Reclamation. Another had been a licensed dentist, and a fifth a successful businessman. Perhaps the Teton had a little more romantic appeal than some of the national forests. Jackson Hole was the early-day rendezvous of murderers and horse thieves, and Teton National Forest adjoins Yellowstone National Park. It was a mecca for elk hunters, many coming from faraway states. Some wanted to stay, and they found a satisfactory berth for themselves with the Forest Service.

The forestry schools began to attract fine groups of students. The graduates were used in various capacities such as on timber cruising (estimating), tree nurseries, planting work, marking and scaling timber, grazing reconnaissance, mapping and survey work. They were involved automatically in fire fighting.

A more technical examination was set up to test these candidates. It was known as the Forest Assistant examination and was usually given near the forestry schools. The one I took in Minnesota was given on the top floor of the post office building in St. Paul. The same Civil Service examiner was conducting eight tests including one for matrons at Indian agencies, one for teachers in the Philippines, another for stenographers. In the teachers' tests the examiner read long lists of words for spelling ability. The stenographers' test included dictation at various speeds. The different tests were announced and started one after the other as the examiner got around to it. Terminal time was announced at the end of each examination. Typewriters were pounding away. It didn't help matters that the desks were ordinary school desks, with the person ahead of you sitting on the seat attached to your desk. And the desks were not attached to the floor by bolt or screw. Imagine a fidgety would-be school ma'am jiggling around on the front half of your writing desk!

Our examination was set up for two days of seven hours each. The first day it seemed almost impossible to concentrate. The second day, by some kind of superhuman strength, everything around me was blotted out. I wrote for the full seven hours without once leaving the seat, totally oblivious of the hubbub around me.

The last Forest Ranger examination was held in 1929. By that time the ranger jobs were becoming less a policing and protection job and more was required in the way of timber management and range management. The Forest Assistant job was changed, in name at least, to Junior Forester, and a Junior Range Examiner job was added. Examinations were held for each. New graduates of the forest schools were appointed from these registers and were assigned to work comparable to the old Forest Assistant places. After a breaking-in period of one or two years, these men were assigned as assistant rangers on the heavier districts. From this group the vacancies in District Ranger places were filled. This meant the man in charge of a ranger district was first a school-trained forester or range manager who had had a good breaking-in including assisting in the management of a unit. An important further policy was established. All Assistant Supervisor and Forest Supervisor positions would be filled only following a successful period as District Ranger. The practice is still followed.

*Forestry Courses**

Various forest schools offer different courses. Forest Management is the usual basic course, with Forest Engineering and Forest Products courses offered either as major courses or as ones supplemental to Forest Management.

---

* See Appendix for list of accredited forestry schools.

Courtesy U.S. Forest Service
A LOOKOUT COOKING BREAKFAST
Another subject not taught in the forestry schools. It's a kindergarten subject for the forester. Firefinder and map on the right. Here cooking, sleeping, and working can double into one small room.

The Forest Management course emphasizes the technical and administrative measures that aim to produce the greatest values from all forest resources.

The Forest Engineering course is designed to prepare men to deal with logging operations and the movement of timber from the woods to the mills.

Forest Products courses include product development, business, and sciences such as wood chemistry and pulp and paper chemistry.

Most schools now offer courses in Forest Recreation.

Range Management may be treated as a separate program, or courses may be offered for students who are majoring in Forest Management.

Through graduate schools it is possible to secure Master and Doctor of Philosophy degrees in Forestry. Those who plan to teach Forestry or go into research work find it advantageous to take graduate work.*

Two Deans

In August, 1950, Arnold Standing, Assistant Regional Forester, in charge of the Division of Personnel Management in the Portland office, spent a week on the Whitman Forest. He had as guests Gordon Markworth, Dean of Forestry at the University of Washington, and Paul Dunn, Dean of Forestry at Oregon State University.

The visitors wanted to see the Eagle Cap Wilderness Area and the high Wallowas close up. Ranger Harold Dahl of the Eagle District and I assembled saddle and pack animals, necessary gear, and groceries at the Ideal Guard Station on East Eagle Creek. Next morning, early, we headed up the East Eagle trail. Eagle Cap, elevation 9,675 feet, was in view all the way up the canyon. It was noon before we reached the saddle close under Eagle Cap and got a look at the Lakes Basin.

---

* See Appendix for courses and brief explanation of subjects.

Here the trail was blocked for three hundred yards with snow from the previous winter. It was steep as the roof of grandpa's big red barn. Everyone dismounted and, cautiously testing the footing, we moved across the big drift without mishap. No air holes were encountered and an inch of melt on the surface saved it from being too slippery.

Two of the prominent peaks are Matterhorn, 10,004 feet, and Sacajawea, 10,033 feet. China Cap, somewhat out of range, is named for the brown lava rock sitting atop its gray granite base. We saw Upper, Mirror, and Moccasin lakes, and camped at Glacier Lake. Before the night was over, we agreed the lake was well named.

Next morning we passed Prospect, Frazier, and Little Frazier lakes, and topped out at Hawkins Pass. The look back over the Lakes Basin couldn't be compared with the panorama of mountains and canyons before us. The visibility was perfect. Besides seeing much of eastern Oregon we could view a big chunk of Idaho.

Our eyes full of looking, we headed down the trail in the South Fork of the Imnaha River. Only a few hardy citizens have seen this part of America. Ranger Dahl volunteered to drop behind and catch some trout. We kept wheeling along, for a car was to meet us at the end of the road at the Indian Crossing. The last five miles was almost too much, but nobody would have quit had it killed him. The two deans really suffered. They would dismount and stiff-legged-ly stagger along until they limbered up. By that time they would decide maybe riding was less misery than walking and they'd crawl back on. Another half mile and they'd decide the original idea must be best after all, and they'd dismount. We'd watch the mile signs and swear they *couldn't* be right. But both deans proved to have what it took. The last quarter mile they agreed they'd go in riding, heads up, shoulders back, and whistling as though they never had heard of a blister. And they did. The fisherman caught up with us and had a fine string of trout.

Herb Hunt, from the Baker office, was at the crossing,

waiting for us with the car. But before parting we had a glorious fish feed. The visitors went out by car to Enterprise and the following morning Herb, Harold, and I started back with the horse outfit.

Training on the Job

When a man is graduated from a medical college he has to serve an extended internship in a hospital. It is somewhat similar with a forester. Many forestry students take summer jobs as trail workers, lookouts, fire fighters, or as chainmen on survey jobs. Experience and practice in the doing of the common, everyday little things mean the difference between a greenhorn and an experienced woodsman forester. I once saw a pretty high-up forest official take a grub hoe and, finding it was loose on the handle, grasp it firmly and hit the end of the handle sharply on a nearby stump. Naturally the head of the hoe slipped down the handle and gave him quite a rap on the knuckles and a painful pinch between the iron and the wood.

A forest school graduate, going on a job requiring a saddle horse, undertook to bridle a big, long-legged horse. He stood squarely in front of the horse with one hand on each side of the bit. The horse kept raising his head as the bit was applied to his tightly clenched teeth and was shortly beyond reach. A packer nearby said, "Let me give you a hand." With the top of the head stall in his right hand and bit in the left, he slipped the bridle over the horse's nose, right hand to the top of his head, left hand to his mouth, and the bridle was in place in less time than it takes to tell.

The point is you don't learn these things in college. Some of the farm boys who have hunted the milk cows in woods pasture, or loggers' boys from timbered communities, or even Boy Scouts who lived near the mountains and forests, have quite an advantage. And many of the forestry colleges are far from the forests.

I wasn't like the buckaroo who said he was born in the

saddle. With a gunnysack for a saddle, I'd ridden old Maggie up and down the corn rows, my older brother Rob at the handles of a walking cultivator. I rode lead horse on the farm grain binder. But a lot of things were yet to be learned about horses unless I was going to go afoot.

Photo by Byron Brinton, Baker, Ore.

A FUTURE RANGE EXAMINER

Many of the early range managers and range policy makers were ex-stockmen. Now they are technical college graduates. But many have hunted cows in the brush, herded sheep, or lost their hats as this boy has.

Early in my first summer in the West, Supervisor McCoy
took me with him to the Sublett District to help survey a
prospective homestead. Mac drove a team of horses, then
rode one of them when we left the station. Ranger Bert
Mahoney not only loaned me his good horse, Brownie, but
put his own saddle on him for me. He rode with only a big
blanket and a surcingle. Mac and Bert mounted and started
out. I mounted, but Brownie didn't like my farm-boy way
of steering. He gave me my first lesson in how a well-neck-
reined horse responds. But he didn't show me up in front of
the old-timers.

At the Bostetter Station, a summer station on the Cassia
Division, Ranger Clarence Nelson and Simon Kinney, his
assistant, had a real horse outfit. Between them they had
nineteen saddle horses, mostly broncs, some being broken for
the stockmen. One of the men would get breakfast while the
other took the wrango horse and rounded up and corralled
the bunch. After breakfast they would pick out the snortiest
two in the corral, rope them, saddle up, and often, after a
little exhibition, ride off for the day. Sometimes the cavvie
was brought in again at noon and fresh mounts taken out
for the rest of the day. A wrango horse was always kept in
the corral.

I'll always be grateful to Sime for showing me the proper
way to mount. Like many new riders, I would grasp the
saddle horn with my left hand, put my right hand on the
cantle and then try to get my right leg over the horse and
my seat into the saddle. My right arm was in the way and I
had to let go of the cantle to get my leg over. I could easily
lose my balance. Sime demonstrated how to take the reins
in your left hand, leaving no slack so you can control your
horse, left hand on horse's neck a foot or so from the saddle.
After putting foot in stirrup, grasp *horn* with right hand.
Then with a slight spring, a pull on the horn and a swing
of the right leg, you're in the saddle, well balanced and in
control of the horse all the time. And Sime, like Brownie,
taught me with no grinning spectators.

Some years later, on a trip to the Wyoming Forest, we started a field trip from Fred Graham's station. He pointed out a black horse in the corral with seven or eight others, saying I could ride Blackie and a spare lariat was by the gate. So I coiled the rope, made a good loop and, as Blackie made a run for it, I made one throw. You know, that silly old saddler stuck his head right in the loop, and when the rope came tight around his neck, he moved right up to me ready for the bridle. To this day Fred doesn't know I might have tried a dozen times and missed every throw.

Probably all bosses aim to give training to their helpers and that is good. Another way to learn is to study your bosses and attempt to profit from them. My first two supervisors were most helpful in both respects. William McCoy was an old-timer. He had been a miner. He was a diplomat, was easygoing, never seemed to have any worries, and never was in hot water. There was a lot of grazing trespass on the forest and some of his men were not up to par. He contracted pneumonia and died before I had worked for him a full year.

Walter Campbell, Deputy Supervisor on the Weiser Forest, was selected to take McCoy's place. In many ways he was the opposite of Mac. He had a world of energy and accomplished a lot in a day. He was a great admirer of Teddy Roosevelt, after whom I thought he tried to pattern. He tackled the stock trespass with vigor and that didn't set so well with many of the stockmen. He dissolved partnership with several of the rangers. He had trouble in his office. He didn't always get backing from the Regional Office. I realized that if a person could adopt the strong points of each of these two men and at the same time avoid their shortcomings, he could be a real success.

The Weiser Forest had more fire guards and assistant rangers than the Minidoka. So when a vacancy occurred on the Minidoka, a Weiser boy was selected to take the place. Before long there were three old Minidoka rangers and three new ex-Weiser men on the one forest. In my job I worked with all of them and was considered, I'm sure, a neutral party. Anyway, I heard both sides. So far as I could see Walt was not partial in the least. But human nature being what it is, it was natural to develop the feeling of "teacher's pet" on one side, and wondering when the next shoe was going to drop on the other side. So! Lesson No. 2: I resolved never to drag along old associates to a new job. And that plan was followed in every move I made. But just a minute here, farm boy, maybe nobody would have moved with you if he had been invited!

In these earlier times, the annual grazing applications were taken in the Supervisor's office. This might be the only time in the year some of the smaller owners were contacted by that officer. Also it was useful and sometimes necessary to refer to the individual's card record for the past use, permanent or temporary numbers, and other data. But to get the card out of the file it was necessary to know the permittee's name. One officer had a lot of trouble on this score and he developed several ruses. He might say, "Let's see, what are your initials?" If the answer was "A. X.," he was still stuck. Sometimes he came out in another room, closed the door, and inquired of others as to his visitor's name. Other times he'd ask the man to sign the application only partly filled out, then finish after he got the card. Or he'd ask the applicant how he spelled his name. That was likely to make his face red if the man's name was Black or White, or if the man said, "Oh! with an 'e' as usual."

Glen Smith, in charge of Range Management in the Missoula office, went into the Missoula Mercantile Company hardware department. A young clerk, smiling effusively, approached him. "What can I do for you today? How are you? How are all the family?" Glen told him he needed fifteen cents worth of shingle nails. As the purchase was about ready for him, foxy Glen began fumbling in all his pockets and finally told the clerk he must have forgotten to bring his purse and could he charge it. "Oh, surely," said the clerk. Then he began writing out the charge slip and queried,

"Let's see, how do your spell your name?" Glen said, "S-M-I-T-H."

Lesson No. 3: It's always better to come right out and admit your memory is failing.

Diversity is one of the reasons why a forester's work is interesting. But it calls for a constant guard against becoming a Jack of all trades and master of none. If you're fighting fire, study fire action, see how older heads perform,

Courtesy U.S. Forest Service
A SUMMER JOB AS A SMOKE JUMPER MAY APPEAL
Other people it makes a little nervous. Aerial work is expanding rapidly. In the picture, the faint streamline in the canyon below is the target area.

try to use your head, adopt the best methods known to you, strive to become an expert. If your job is talking to a roomful of schoolchildren, plan ahead what you're going to say and do. See how older heads do the job.

If you are marking timber to be cut and have a more experienced man working with you, or are visited by an inspector or supervisor, ask questions, observe his work, correct your errors, study the problems, improve with doing. It's better to forget a little than to be mortally afraid of learning too much.

Meetings and Conferences

In common with many professions, businesses, and industries, the United States Forest Service holds many meetings, large and small. Usually each meeting has several purposes, such as training of personnel, exchange of ideas and experiences, formation of new policies, and dissemination of new policies and instructions.

I attended the Service-wide Fire Conference at Mather Field, California, in 1920. Colonel W. B. Greeley, Chief Forester, was in charge. Roy Headley, Chief of Operation, and Major Evan Kelly, in charge of Fire Control, both from Washington, D.C., were assistants. All the Regional Foresters and Fire Research men were attending, as well as the regional fire men, one Forest Supervisor and one District Ranger from each region. The meeting lasted for two weeks. We were assigned a barracks at the air base, ate with the military, and had meeting rooms as needed. Much of our work was done in committees whose findings were presented to the whole group for discussion, revision, or adoption. All phases of forest fire work were analysed, practices reviewed, changes considered, improvements planned. An intensive and serious effort was put forth. Everyone worked hard, with evening sessions the rule. A few diversions were provided. Each day, for half an hour before the evening meal, we had football kicking and catching practice, Colonel Greeley right out with the boys. Everyone was taken up for an airplane ride in a big Ford plane. It seemed big then, anyway. It was the first time I'd looked down on timber from the air.

One morning a noncom entered several of our meeting rooms with a private from the guardhouse. He gruffly ordered the private here and there, directing him to pick up this scrap of paper or that cigarette butt. There were some whispered adverse comments about the quality of discipline in the Air Corps. That evening at mealtime word got around that a man had escaped from the guardhouse and a search party had been organized. We had an evening session that night. About nine o'clock the door burst open, the same disciplined private of the morning charged in, dodging here and there. He was hotly pursued by a sergeant, shooting as he ran. Everyone in the room was dumbfounded and frozen to his chair. Roy Headley was the first to come to and he shouted, "Catch him, don't kill him! Catch him, don't kill him." By this time the escapee was just passing Headley, who made a flying tackle, catching him with both arms around the thighs and bringing him to the floor. Whereupon both military men laughed heartily and left in the most friendly way. They had taken a bunch of foresters.

A small group meeting of some of the nearby foresters was held in Boise, Idaho. After a full day we went out to the Nat to relax. Some went into the plunge, others sat alongside the pool. One ambitious swimmer, a young woman of unusual proportions, attracted our attention. She would throw a leg and an arm over the log and struggle up on top of the floating stick, only to have the log roll, throwing her off. Finally Johnnie Raphael, a staff man in the Ogden office, reached over and tapped Regional Forester Dick Rutledge on the knee and said helpfully, "Say, boss, your pipe is out."

Ranger meetings of two or three neighboring forest areas were conducted even fifty years ago. Discussion followed after assigned presentations. Important changes in policy were explained by someone from the Regional Office. These meetings got neighbors acquainted, had educational value, and built up esprit de corps for which the organization has been noted.

Such a meeting was at Blackfoot, Idaho, in 1916. Standard uniforms had not been adopted then, but were under discussion. Robert Clabby had a pair of neat, well-trained mules that he drove to a buckboard over the mountain roads on his district. Someone said that a uniform would identify a man as a forest officer. Robert explained that he didn't need any clothes to identify him, that folks knew who *he* was as far away as they could see his mules and they would announce, "Here comes Clabby and his mules."

Another ranger meeting at Challis, Idaho, in 1920, was attended by men from Salmon, Lemhi, and Challis forests. One of the boys from the Salmon got into an argument with one of the Challis rangers. The discussion, somewhat rough, with a display of personal feelings, was choked off by the chairman. That evening a meeting was called and shortly the two men involved in the afternoon disagreement resumed the squabble. Almost at once one called the other a so-and-so and each whipped out a six-shooter and began firing at a range of ten feet. The lights went out. Pandemonium reigned. We were meeting in a vacant store building with shelving along one side. The lights suddenly came back on and what a sight! Normally dignified forest officers were clinging to the shelves, some with heads pressed to the ceiling. Hog piles of four or five human forms were in the two back corners. Lone eagles were flat on their stomachs, faces pressed to the floor in an effort to avoid the gunfire. Others had sought shelter under heavy tables. No pools of blood appeared. Sooner or later it dawned on everyone that we had been hoaxed. David Laing, supervisor of the Challis, who manned the light switch, and the two principals were the drama team, all rehearsed for the play.

Details and Transfers

A detail to some other job for a week or a month was excellent training and it enabled somebody higher up the line

Photo by Jim Anderson, Oregon Museum of Science and Industry
HENRY TONSETH, RANGER FOR MANY YEARS ON THE
FORT ROCK RANGER DISTRICT
Portrait of a well-liked, hardworking, and dedicated forest ranger. The background
is a big ponderosa pine.

to get more of his work done. Sometimes it seemed that was the main purpose. Of course, after the detail, one had to go home and catch up on his regular job.

From as early as 1910 it has been regular practice to detail foresters to the larger going fires. They made better help than temporary pickup men. They were valuable as scouts, foremen, sector bosses, timekeepers or camp managers, depending on their experience. Or if they hadn't been baptized by fire they still fitted into a big organization job and were welcome assistants. They profited, too. Not only did they get into the fire action—they often had opportunity to see new timber types, new geography, and practices new to them.

Several of the old-time forest assistants spent more than one winter in the "Bull Pen" on the top floor of the old Forest Service building on lower Twenty-fourth Street in Ogden. Some had been on timber survey during the summer while others had been on more general work on some forest. They were assigned to drafting, compilations, and detail jobs. Jim Hull was working on a timber type map of the Teton National Forest, Wyoming. He upset a bottle of light-green ink (the symbol for lodgepole pine) making a big island in the middle of Jackson Lake. He was the lake planting specialist for lodgepole pine thereafter.

Jim invited Roland Stretch, another forest assistant, and me, to go as his guests to hear Harry Lauder, who was making a one-night appearance in Ogden. As soon as the show started it was apparent we were at a vaudeville show instead. I've never quite forgiven Jim for my having missed the only opportunity I ever had to hear Lauder.

Another winter I was assigned to handle a ranger correspondence course. The job included rating all the papers sent in by over one hundred "students." A wide range of forestry subjects was covered. I had a big advantage over the participants because teacher could look up the answers. It was great training for the instructor. Someone above me probably saw I needed that practice.

The most frequent use of the detail as a training tool was to bring a Forest Supervisor or an assistant into the Regional Office for some particular job. In later years it became common practice to select an Assistant Regional Forester, for example, to go to the Chief's office in Washington, D.C., for a detail of two or three months. This broadened his outlook and gave the higher-ups a chance to observe him in action at close range.

Outright transfers commonly are used for training purposes, too. Men are deliberately moved from the west side to the east side or vice versa in Washington and Oregon because of the big difference in conditions. Or men are moved interregionally to try them out in wholly new surroundings with new associates.

Someone said the old army game used to be to recommend highly for transfer to a good vacancy a person who wasn't appreciated in his own outfit. It's only hearsay with me. But to head off schemes of this kind the Forest Service early adopted what was known as the "Treaty of Denver." It was a voluntary agreement that no supervisory officer would use the transfer route to avoid handling a personnel problem. If a transfer was proposed the receiving officer would first be informed of the shortcomings of the transferee.

Specialists in personnel management are more recent. The aim has been to select, train, and develop better qualified men for work that becomes more complicated as time goes by. It's a far cry from the ride, pack, chop tests. But you can't improve a model without losing things, too. Many remember the old Model T with affection.

ONE FORESTER'S EARLY-DAY TRAINING AND EXPERIENCES

ARNOLD R. STANDING
Retired Assistant Regional Forester

Boys who grow up near a National Forest often start to

work for the Forest, find the work interesting, and make the Forest Service their career. In 1918 before I was through high school I began work on the Caribou on an "improvement crew" for $75 a month. On this crew five of us worked under foreman Arthur Peterson, who later became a capable ranger on that Forest.

We started work in Montpelier Canyon, near the "elbow," building a range fence. We had a farm wagon loaded with materials to build or repair roads, bridges, fences, telephone lines or buildings, plus our bedrolls, grub, and camp equipment. From Montpelier Canyon, the crew worked northward, ending on Fall Creek near the north end of the Forest.

Shortly, I was given another assignment. I helped a newly appointed assistant ranger build stock-watering troughs south of Montpelier Canyon. He had the typical equipment of those days, a team of light horses that were used both for riding and packing and to pull a buckboard. This man carried a Colt .45 automatic pistol and apparently relied on it, rather than diplomacy, to obtain compliance with Forest Service requirements. One time he forced a sheepherder, at gunpoint, to travel several miles to a Forest Service telephone. The herder hadn't moved the sheep along the stock driveway as fast as the ranger thought he should. He kept the gun pointed at the herder while he reported the case to Supervisor Charlie Simpson. He asked Supervisor Simpson what he should do and Simpson replied, "What *would* you do? Turn him loose!" Anyway, the man was released during his probationary period on evidence that he was not ranger material.

I met Supervisor Simpson at the Trail Creek Guard Station for another assignment. Charlie and Ranger Jim Bruce had been riding the range; and my first sight of these men, who were to become lifelong friends, was as they came splashing across the Blackfoot River on their fine saddle horses.

Charlie and I headed for Caribou Basin the following morning in his Ford roadster with turtleback replaced by a small box. Along the way, through the small settlements of

Wayan, Gray, and Herman, we met livestock permittees and others who had business with the Supervisor. I was impressed by the importance of Forest Service work and by the sagacity of the Supervisor in dealing with people. Charlie gave me some good career guidance on that trip.

Herman was pretty much a ghost town. The old saloon and several other empty buildings were still standing. I heard several stories about its rough and tough existence, including a man being buried alive up to his neck for some infraction. With Grays Lake's mosquito population, that could have been a terrible ordeal. I can't remember how long he was kept in his "grave."

Ours was the first automobile to travel into Caribou Basin. We had to put rocks and brush in the ruts to clear the high centers.

In Caribou Basin Supervisor Simpson, Ranger Lewis C. Mathews of the Grays Lake District, Ranger Spackman of the Freedom District, and I assisted Dave Shoemaker in one of the first larkspur grubbing projects. Dave, who was a

Courtesy Charles Simpson, Baker, Ore.
AN EARLY-DAY RANGER STATION
Many a forester's bride spent her honeymoon in a home such as this, twenty-five miles from a neighbor and no garage needed.

Range Examiner from the Regional Office at Ogden, Utah, was experimenting with larkspur eradication. I was impressed by the need for it, as we found eight dead cattle in patches of larkspur in Caribou Basin. I believe Dave's investigations were the basis for the bulletin, "Larkspur Eradication on the National Forests." Ten years later I introduced the use of herbicides for larkspur eradication in Region No. 4. A check on the effectiveness of previous eradication showed that in most cases the larkspur became reestablished, but cattle losses prevented for several years more than repaid the cost.

Dave Shoemaker was one of the most capable and admirable forest officers I ever knew. Unfortunately, he died when still a young man while serving in the Southwestern Region of the Forest Service. The story was told that in his youth he and his brother mounted broncos in a corral in Nebraska, opened the gate, rode their bucking broncs in different directions, and didn't see each other again for several months.*

One Sunday I hiked from our camp in Caribou Basin up to the Caribou mine. In the open country at the south end of the basin, range cattle suddenly came running toward me from all directions. It gave me a queer feeling to be encircled by sniffing, staring cattle. There were no trees to climb, so I gathered some throwing rocks and went straight ahead. The cows gave way to let me pass. Later I read in a report by Adolph Murie on losses of range cattle by grizzly bears in Jackson Hole, Wyoming, that cattle sometimes surround bears in this same manner. A bear will suddenly charge at a cow and, before the startled animal can turn and run, the bear will deliver a fatal blow with its paw.

I rejoined the work crew at Grays Lake. This was an interesting area, with a number of "characters" left over from the mining-boom days. One man invited us to spend the evening at his ranch home. Women's clothing was hang-

---

* For other well-known graduates from Nebraska under the tutelage of Dean Bessey and Professor Philips, see Appendix. Those men inspired their students with a zeal for conservation.

ing in the entrance hall. He frequently mentioned his wife and expressed regret that she was not at home that evening. Next day another rancher asked us where we had been. When I told him, he asked if the man's wife was at home. When I said "No," he laughed and said the man never had had a wife.

The crew members formed a hillbilly orchestra. One man played a cornet, another hummed through tissue paper held tightly over a comb, I played a harmonica, and another did a pretty good job of trap drumming, using an assortment of boxes, pans, and kettles for sound effects. We sang and played old tunes such as "Darling Nelly Gray," "Redwing," and "Juanita." We invited the local folks to a concert. People came miles.

One of those who attended was a girl called Tex Heath. She had a beautiful, spirited horse and would ride up and down the road in front of our tent. She became a top rodeo performer in Wyoming. We were told that her father had a large family and guided their destiny by the maxim "The Lord will provide." He took them out on an island in Grays Lake to live. Local people feared the family would starve and insisted that he make other arrangements and help the Lord a little.

About the first of August, all members of the crew but me had to go home to harvest hay, so the work was discontinued. Fortunately for me, at that time an assistant was needed for the cattle herder on Brockman Creek, and I got the job. There were 1,100 cattle ranging on the allotment. The herder was a likable Dane. His method of making coffee was to pour a whole package in the pot and fill with water. This would be boiled and the coffee poured off the top. Then more water would be poured in and reboiled. After several days, a new package of coffee would replace the old. The herder had stomach trouble.

One of the saddle horses in our string had a bad habit of rearing straight up when mounted and then falling over backward. The herder watched me fall off and scramble out of the way twice, and then showed me what to do. He passed each bridle rein around a tree before mounting and thus restrained the horse until he got out of the notion of rearing, and then reached down, unwrapped the reins and was on his way.

In 1919, after graduating from high school, I again headed for the Caribou country. I worked a short time for the Forest Service in Caribou Basin as administrative guard, $90.00 per month, but I was offered and accepted a higher paying job at the Caribou Mine. Ranger Mathews didn't mind my leaving, and I needed all the money I could earn.

After the mining job played out, a sheep permittee named Lewis White needed a sheepherder just then, and I was glad to get the job. Sheepherders' wages were relatively high, and Supervisor Simpson had advised me to learn all I could about range livestock, as I had decided to become a range examiner in the United States Forest Service. The range allotment was at the head of Jackknife Creek. One camp tender moved two camps. He was a buddy of the other herder, so I seldom saw him except when I needed supplies or wanted to have my camp moved. "Bedding out" and deferred and rotation grazing were not practiced then, or at least we didn't. I had to learn sheepherding by myself but got along O.K. Bears were quite numerous in the area at that time.

I had the use of a dog that knew more about sheepherding than I did. I could guide him as far as he could see me by motions of my arms. If the sheep needed to be turned or brought into camp, I would signal to him. His only fault was that he was deaf. When I would sit down, he would lie down near me. If I walked away, he would stay where he was unless I made an arm motion for him to follow. Sometimes I would forget about him and walk away some distance, when I would hear him howling. He couldn't hear me call, so I had to walk back to where he could see me motion for him to come.

Later on, after attending Montana State University and Utah State University, I received an appointment as Forest

Ranger on the Cache National Forest, April 1, 1923, and as Junior Range Examiner on the Fishlake National Forest, Utah, on June 1, 1924, which started my permanent career with the United States Forest Service. I have always cherished my early experiences on the Caribou.

CHAPTER FOUR

FIRE IN THE MOUNTAINS!

Forest Fires

SOME FOREST FIRES HAVE BEEN MAJOR NATIONAL DISASTERS. One was the Hinckley fire of 1894. Hinckley, Minnesota, is seventy miles southwesterly from Duluth and fifteen miles west from the St. Croix River, the boundary between Minnesota and Wisconsin. It is timbered country, extensively logged prior to the time of the fire. Limbs and tops of trees, cull logs, and windfalls were everywhere, and small smouldering fires burned at will. Following several weeks of dry weather and low humidity, the wind on September 1 fanned the small fires into hot ones that ran together and spread a wall of flame. It produced its own draft, giving forth a terrific roar. An estimated five hundred people barely escaped from Hinckley in a burning train, across burning bridges. The town was completely consumed as were smaller villages both north and south of Hinckley. Many persons were severely burned. All were left destitute, without homes, food, or clothing. The number who died that day in the Hinckley fire was placed at 418.

I remember well that fire even though I was a preschooler. My parents, living little more than one hundred miles south of the fire, filled an apple barrel with clothing and nonperishable food such as beans, dried corn, dried fruit, sugar, flour, and home-cured ham. Each of four youngsters in our family divvied up toys and playthings for the homeless children of Hinckley.

Another national calamity was the widespread northern Idaho-western Montana fire of 1910. Although I became intimately acquainted with that area later, my 1910 information was limited to what I read in the papers about the loss of life and destruction of timber, villages, and wildlife.

In 1911 as a forest school student I got a job as a fire guard on the Superior National Forest in northern Minnesota and took the night train out of Minneapolis for the North Woods. The train went through Hinckley just at dawn. Not an evergreen was visible for miles. Black snags were everywhere. Mercifully the charred earth was clothed with fire cherry. Recollections of stories of the sufferings of the Hinckley people were vivid. And the disasters of North Idaho just the year before didn't ease the picture of that horrible graveyard scene. I, a poor, inexperienced youth, was going into that

Courtesy U.S. Forest Service

LIGHTNING IN THE SKY

Lightning causes many forest fires but at the same time it flashes an alarm. The smoker and camper may cause fires but neither one gives an alarm.

tinderbox to work as a fire guard. That's when the boy almost got separated from the men.

Causes of Fires

Very early some farseeing top forester decided upon a detailed record of every fire. A small but adequate fireman form was devised and from this and one or more lookout reports, the individual report was prepared.

One of the important sections was the "Cause of Fire." Under this for checking were: "Lightning, Campfire, Smoking, Slash Burning, Railroads, Incendiary, Miscellaneous."

Also the kind of person, as: "Stockman, Camper, Miner, Logger, Rancher, Hunter, Fisherman, Other."

With this record, an analysis could be made for a ranger district, a forest, or by years to determine action. In some groups, efforts could be pinpointed. With recreationists more scatter-gun educational efforts were necessary.

Lightning fires in most areas far outnumber those caused by man. They can usually be identified by the presence of splintered wood or the telltale groove spiralling down the bark of a green tree. In the Kaibab country hundreds of trees can be found with lightning marks but no signs of fire. Scarcity of duff, limbs, and young growth around the base made fires hard to start and rainfall usually came with the electric storm. I've seen fires start in the treetop and no fire develop on the ground. Lightning does peculiar things. In one case of a big yellow pine, the lower two thirds of the trunk had exploded and big chunks were thrown for rods. The stump exploded and left a hole big enough to bury a horse, while the top third had dropped down intact and lay across the stump hole.

Jackman recalls that while driving from Redmond to Prineville, Oregon, he saw, a hundred yards in front of him, a big juniper explode exactly as though a dynamite charge had been set off. Tree chunks flew in all directions. He walked all around the burning tree and stepped off the most distant piece of wood he could find on all sides. They were scattered fairly evenly in a circle reaching outward from the tree almost 150 feet each way.

Probably the explosion is due to almost instantaneous generation of steam in the center of the tree. But most strikes do not have that effect. They are more likely to wind down the trunk from the top to the ground.

Not many fires are caused by spontaneous combustion. I knew of one ususual cause. Someone had cleaned out a cabin and left a pile of junk outside on the ground. Included were many bottles, some empty and others with various contents, such as drugs. The absence of tracks or signs of the recent presence of man convinced us that either the sun shining on a bottle caused the blaze, or some chemical was to blame.

Coal-burning railroad locomotives started many fires, and even diesel engines can do it. But special spark arresters were provided, rights of way were cleaned, speeders followed trains, and section crews were available to go to the fires. With Harry Gisborne, a fire studies man, I once had a wild ride on top of a night freight from Missoula to Butte, Montana. We had some shallow boxes filled with materials to check ignition by sparks. We checked how far from the tracks sparks were going, what percent were still glowing when they hit the ground, and the effect of steepness of grade upon spark throwing. In fact we were comparable to the sheepman in South Idaho. He could, so they said, count a band of sheep and tell you the number of ewes, the number of wethers, the number of lambs, how many bells, how many blacks, and how many chambermaids in the Bannock Hotel.

Incendiary fires were rare in our regions. Some cases were suspicious.

In one case three separate fires were started and evidence (burned matches and unburned kindling) was proof of intent. A chap showed up as a fire fighter who had been seen coming from the locality, and later he admitted starting the fires to get work.

No place on the report is provided for "Unknown." The person on the ground is better able to arrive at the cause than anyone else and an educated guess is preferable to a catchall "Unknown." If the fire had started near a stream, if no storms had occurred, and if no sign of a lightning strike could be found, it was logical to decide that it was a smoker fire started by a fisherman. If it had occurred beside a trail during deer season and no evidence of a campfire showed, it would be listed as smoker-hunter caused.

Fire Prevention

To impress the public with the need for greater care with fire in the forests, American Forest Week (Fire Prevention Week) was announced. A concentrated effort was made to reach the maximum number of persons. This was started before TV and before radio was common. Posters were displayed in windows. Articles were run in newspapers, and every school, rural or urban, was visited and talks given, illustrated with colored slides. Since most of the rural schools did not have electricity then, a portable generator was taken along to operate the balopticon. Enough army blankets were provided to cover the windows in the biggest schoolhouse because the programs were held during school hours. Many a romance between forest rangers and school ma'ams started as a result of those Fire Prevention Week programs, and lots of marriages followed.

On one of these visits to schools in Montana, Red Stewart, an assistant in the Regional Office, went along as a helper and was inveigled into doing the talking while we showed a set of wildlife pictures. It was his first attempt and he wasn't prepared to give much of a story as the pictures changed. His contribution went about like this. "This is a buffalo." "This is a bull moose." "This is an elk." Before Red could give his tabloid explanation when a black bear showed on the screen, a lad in the audience piped up, "This is a bear." It almost stopped the balopticon.

Eventually Smoky the Bear grew up and he seems to have really caught on and is still popular and eye-catching today.

Campfire permits were issued by the hundreds of thousands. The idea was not to have a folded slip of yellow paper in every camper's pocket but to cause him to think enough about his fire to go out of his way to procure the permit and to give the issuing person an opportunity to pass out a few well-chosen words of caution. The permit system no doubt prevented many a left campfire. But the feeling grew that the careful campers got the permits and the careless ones didn't. The permits were dropped in some regions. Gatemen were tried at the entrance of some districts; each party was stopped and cautioned about fires. Fire prevention guards were used to travel popular roads and visit campgrounds,

Courtesy Charles Simpson, Baker, Ore.

AN EARLY-DAY LOOKOUT

This homemade map board on a stump stands exposed to flying ants and to all the elements.

talk with campers, and demonstrate care with campfires. Other areas were closed to entry during critical periods, a locked gate or a barrier placed to prevent entry.

But analysis of the fire records disclosed that smokers were more of a problem than campers. Smoking was prohibited except at camp (within certain areas and during designated periods). Some groups of travelers would decide at most any hour to announce a "camp" and stop and light up. That was before the day of the coffee break. We met a sheepherder on the trail and as he saw us approaching he cupped his left hand, filled it with tobacco juice, and dunked a glowing cigarette.

The "No Smoking" rule was debated in many Forest Service meetings. Eventually "No smoking while traveling" was adopted as the best solution. The goal is to cause people to become fire conscious and thereby reduce the number of fires.

Detection

Before a fire can be put out it must be "spotted" and located. Many fires were found by cooperative individuals, others by Forest Service employees traveling roads and trails and visiting patrol points. In the quite level pinelands of Minnesota we would climb trees growing on slight rises or make a pole ladder to get on top of a house-sized rock.

Later a few prominent peaks were selected and used as lookout points. High Point Lookout on the Targhee Forest in Idaho, the highest point on the forest, was located sort of at the forks of the overalls. One large leg of the forest extended south and another extended west. John McClaren, Fire Chief in the Denver Region, and I made an analysis of fires on the Targhee. The local foresters looked on High Point as a primary lookout covering almost the entire forest, and thought of it as a valuable part of the fire organization. The analysis, surprisingly, showed that High Point had made no first discoveries during the years it had been occupied. Why not? Fog and clouds between the lookout and the fire; haze and dust blowing in from the big valley to the west; intervening ridges; too great a distance. Anyone or everyone can see a fifty-acre fire—the important thing is to spot a smoking campfire or a single snag afire from a lightning strike. It doesn't help much to have a friend phone that your house is afire if you are already moving out the furniture. First alarms are what count.

While on a visit to the Salmon Forest in Idaho, Supervisor Scribner and I rode up to old Bluenose Lookout from the back side. No one was on the peak. Scrib rang the lookout's ring and the lookout man answered. He was at his camp a quarter mile below and on water. After inquiry as to visibility and presence of smoke and getting suitable answers, Scrib requested that the lookout watch a certain ridge across the river for signs of fire and call the station in fifteen minutes. In five minutes a surprised lookout appeared at the point. He couldn't say a word and it wasn't because of his speedy climb up the trail. He packed his bag.

About this period the policy was developing to get living quarters of all lookouts right on the peaks. Some thought it would be too dangerous, but it proved more effective. The lookout was observing most of his waking hours. It meant backpacking his water to the peak, but in some cases tanks or cisterns were provided to store snow water, or water was brought up by pack string.

At first two men were kept on each lookout. In case of a fire one would stay on the peak and the other would go to the fire. By now we were making what were known as "seen area maps" for each observation point. Areas seen directly from the peak were colored as "seen." The areas on the yonder side of the ridges or areas obscured by an intervening ridge were left blank. Only areas within twelve miles were credited. These mapping jobs showed up a lot of blind areas, calling for more lookouts. Other points were selected to cover the largest percentage of blind areas. On the Coeur d'Alene Forest, a high-value, high-hazard white-pine country, two men were stationed on almost all lookouts. It was quite a jolt to split up those two-man stations. In addition to increased detection coverage, the men had only half as far to travel to fires. The single man became a lookout-fireman and his peak would be left uncovered while he went to a fire. It was a case of putting out a known fire in preference to looking for one that might not occur.

The change from full-time lookout to lookout-fireman brought about employment changes. One good lookout had a peg leg and couldn't serve as a lookout-fireman. The change eliminated the employment of girls and women unless a primary lookout with lookout duties only was available. The opportunity for man and wife teams was improved. The practice developed of paying the man by the month, paying the wife by the hour when he was away on fires and she substituted as lookout. In instances the wife was a more alert lookout than her husband. Many of the jobs were filled by forest-school boys. They came recommended by their schools and knew the institutions would have a performance report, an advantage all around. The boys had summer training; and their school had a check upon their performance.

All lookout stations were provided with map boards and one of several sighting devices. The Pacific Northwest Region devised the Osborne firefinder, with front and rear sights to fix direction and a device to determine and show on the map the distance from the peak—the two factors giving the location of the fire. The Northern Rocky Mountain Region used a lower-cost alidade. It revolved on a pin in the center of map and had one slot sight and a hair sight. This was accurate as to direction but distance had to be estimated. If shots were available from two lookout points the dispatcher could pinpoint the location by crossed strings on his map.

Some men found the life of a one-man lookout too lonesome. Ranger Ed MacKay, on the big Powell District, used to tell his men at training camp that when he was a look-

RANGERS LEARN TO USE WHAT IS AT HAND
A ladder made on the job provided access to a natural lookout point.

out he kept busy. He'd build a table today and maybe re-build it tomorrow. He'd dig a garbage pit and maybe change its location several times. One lookout I knew prided him-self upon his pies. One day he had some visitors and he served delicious fresh huckleberry pie. One of the ladies was real complimentary, not only about the berries and quality of the crust but the artistic scallops around the edge. She asked how he made such a beautiful pattern and he told her that was his secret. But she insisted and he finally said, "I use my upper plate."

I sat out a severe electric storm on Graves Peak, elevation 8,287 feet, on the East Selway one night. The combination living quarters and observatory had lightning protection. This consisted of four tall poles forming a square with a No. 4 copper wire connecting each pole at the top and two diag-onal wires from opposite corners, making a sort of net above the building. Ground wires went down each pole, and about two feet apart on the overhead wires were short wires at-tached to the main wires so that an end projected upward about four inches. While the storm cloud enveloped the peak, blue flames flowed from those fifty or sixty points. The visible phenomena were accompanied by the noise of frying bacon magnified about twenty times. The combi-nation was eerie but no danger was involved. Stoves, steel cots, and other metal objects were provided with ground wires and chairs and tables had glass insulators on the legs.

In such cases the telephone may ring steadily but the switch is normally pulled. One who hasn't seen an electric storm on a high mountain peak has missed one of nature's brightest and most active shows. An angry rain- or hailstorm may accompany the display.

On the Coeur d'Alene one year we had a corpulent eigh-teen-year-old on one of the peaks. He had a six-shooter and entertained himself by target practice. He had no holster, using his back overall pocket. He'd practice drawing and shooting. On one try he shot a bullet through the fatty part of his rump. This did not pass unnoticed. The question was, "What happened? Were you too slow on the draw or too fast on the trigger."

As more of the secondary or lookout-fireman stations were selected, often at lower levels, it was necessary to do extensive clearing or to build a tower with a lookout house on top. Thirty- or forty-foot wooden towers were common. Some steel towers even one hundred feet high were built with the living quarters on top. Some of the wives thought that was quite a way from the woodpile.

Since World War II it has been common to use planes following severe electric storms to supplement the lookouts and cover blind spots. In 1964 several forests tried substituting planes for all lookouts. The whole system of fire protection and control is constantly changing as new knowledge and equipment come along.

Communication

There were no telephone lines in the forests when the national forests were proclaimed. Progress was slow at first and consisted primarily of getting a line from the ranger headquarters to the nearest commercial line or switchboard. The stations were usually located at some central spot on the district where there was plenty of the essentials: "wood, water, and horse feed." Later, as the families required schooling and more business was delegated to rangers, adequate headquarters were provided in towns or villages, or at least at places accessible to the public, and habitable the year long. Trial and error dictated the type of telephone lines. Some wide right-of-ways were cleared through the timber. In other cases short crossarms were spiked on trees and two wires stretched tight as was the custom on commercial lines. But snags topple in the forest and live trees are thrown by the wind. Broken lines were too frequent. Even heavy swaying of the trees broke the wires.

Shortly a grounded or one-wire line came into use. Each end of the line is grounded, completing the circuit, instead of a two-wire metallic circuit. Instead of a solid tie, split tree insulators were used. An oval kind two and one-half inches long was adopted. The two halves, just alike, fit together astride the line wire. You get the same effect by extending both hands and putting the palms together loosely with a wire between. A short piece of wire of lighter gauge than the line wire is wrapped twice around the insulator in the grooves provided. The ends are twisted and two short hooks are made that fasten into a large staple driven into the line tree. With lots of slack in the line, if a tree falls across it the wire pulls through the insulators taking slack from both directions. The tree can fall without breaking the line. If too many trees fall or there isn't enough slack, the hook will pull out of the staple or break before the line wire will break. Only some brushing out is needed to hang the wire. Repair is quick and inexpensive. In case of a line not used in winter, the spring maintenance is a mere fraction of what it would have been with a solid-tie two-wire line. With this system it was possible to provide communication to new lookouts and fireman stations as the fire organization expanded. The draping lines didn't look like much to a telephone man but they worked. In rare instances deer or elk got their antlers twisted and tangled up in a wire and died fighting. One fireman was sent out to repair a tree line. Upon his return he reported he had found the break but couldn't get the ends together so had used his halter rope to make the connection. The station man said, "Does it work?" The fireman replied, "Oh, yes, it works if you talk loud enough."

Substitutions were tried. A few heliographs were put out. Two lookouts can communicate by using the Morse code. The heliograph includes a mirror and a shutter and requires sunshine. Opening and closing the shutter provide dots and dashes. An ordinary pocket mirror has been used by a man on a fire to flash back his beam of light to a lookout point. That shows his own location. Or two men using hand mirrors can signal to each other on a sort of one if by land, two if by sea basis. Workers in a fog-infested area would better forget about these schemes.

Homing or carrier pigeons were tried. One ranger on the Thunder Mountain District in Idaho built a small loft and at Bend, Oregon, homers were experimented with for some time. But it's a one-way trip and although the army found some good use for them, they aren't practicable for day-after-day, close-interval communication in a forest-fire organization. It's a little hard to ask questions back and forth. They are quite popular for competitive racing and races over various courses up to 150 miles or more are common.

My young son Kenneth and I raised and trained homers and had a lot of fun. Training starts when they are young by taking them out from the loft one hundred yards at first and then lengthening the trip. A numbered aluminum leg band is put over a foot of a newly hatched bird and the band is there for life. Message carriers are small aluminum tubes attached to the leg. Messages on tissue paper are rolled

Photo by Bill Bell, Missoula, Montana
PACKING LUMBER FOR A LOOKOUT HOUSE ON THE LOLO
NATIONAL FOREST, 1926
No beginner's job here. Imagine what would happen to this string if they ran into a
nest of yellow jackets!

up and slipped into the tube. I often took three or four homers out on a trip on the forest and turned them loose with messages to the boy.

One of the practices is to take one of a mated pair out on a flight. He will just bust himself to get back to the mate at home. We never knew of a divorce among our mated pairs. They mate for life, same as geese. But if one fails to come home within a week or even less, the mourning period is over.

We had three birds with official flight records of one hundred and fifty miles. Ken bought a pair, second-generation offspring of the Welch Wonder Hen, a speedster that flew from the Bay of Biscay, Spain, to a village in Wales—nonstop flight. To check on the speed, Ken and I set our watches to agree and I took four fliers out to Copper Mountain Lookout. Scaled off on the map, this was just fifteen miles airline from our loft. I tossed up one at a time at ten-minute intervals. Ken checked their return. The fastest made fifteen miles, including getaway and descent, in just fourteen minutes. The second best was fifteen minutes, the third, sixteen minutes. This is sixty miles per hour. The last one, enjoying his freedom, took over an hour. Maybe his compass needed adjustment, or he didn't believe it.

Now that most all forest stations are equipped with radios, with radios in cars and walkie-talkies for firemen and others, these early-day makeshifts are past history. Even miles of telephone line have been replaced by radios.

Putting Them Out

The fireman (earlier called smokechaser) is the boy who goes to the fire and puts it out. The great majority of forest fires can be put out by one trained man if he gets to it soon enough. Any fire is a forest fire, however small. An unattended campfire, a lightning strike in a single tree, a burning stump, a log used to provide a warming fire and left by a hunter—all will qualify. Any fire might become a Hinckley disaster with proper conditions. So! Get to the fire. Have everything you need ready to go. The fireman's pack includes tools, water sack or canteen, food, camp kit, compass, first aid, flashlight, map, pencil and paper, report forms. If he omits one of these, extra time may be used up and a fire may get away that could have been put out. A Number O or ladies' shovel and a Pulaski tool are the usual fireman's working tools. The Pulaski tool is a combination axe and grub hoe. It was named for Ed Pulaski, ranger on the Wallace District of the Coeur d'Alene Forest during the 1910 fire. He had been a blacksmith. He took a double-bitted axe, turned one blade ninety degrees and retempered it. That gave a digging tool. Handmade at first, factories make them now.

Except in the earliest of fire-fighting days it has been standard practice to go to fires night or day. Night travel calls for lights. At first we used a "Palouser." This was a lard bucket with a hole cut in the side, four points turned in and a tallow candle pushed through. As it burned it was pushed farther through. A bail attached to the top and bottom served as a handle.

Carbide lights followed. They were fashioned after miners' lights—were lightweight and gave a fair light. But they had an exposed light and were easily clogged. They required water and a supply of carbide.

Stonebridge lanterns came next. They were pretty things and, being collapsible, were easily carried when not in use. They were four sided, about four inches square and six inches tall, with light metal frame with isinglass windows on the sides. A special wax candle fastened onto the bottom. They didn't give much light.

Old-fashioned coal-oil lanterns were used, especially with small crews. For fireman use there was too much weight, and spilled kerosene was bad.

Electric flashlights came next. Most any kind was used first and those big focusing ones were great. The best arrangement was a headlight with battery box on the belt and cord connection. This freed both hands and the light was thrown where you were looking.

Providing emergency rations was difficult. They must be light, nonperishable, nourishing, and acceptable to the majority. Army K rations were tried. Lots of users were unhappy with them—too concentrated and unpalatable. I knew one chap who said, "I want something to eat," as he threw out the K rations and put in half a loaf of bread and a chunk of boiled beef. Some had lived on them in the army and were fed up. Dehydrated foods were tried. A firefighter is often not on water and he hasn't time to cook. He must eat on the run. A quite acceptable one-day sack ration includes a can of brown bread, small cans of fruit and baked beans, tinned cold meat, coffee, and candy bars.

In the fall of 1913 Nils Eckbo was making planting surveys on the Minidoka Forest. He had brought along some dehydrated foodstuffs and on Thanksgiving Day we decided to put some of it to soak. We were camped at the Frazer Ranger Station (east of Rogerson, Idaho). Soon we had dishpans, stew kettles, water bucket, pitcher, dripping pans, everything in the place filled and running over with cabbage, berries, spinach, string beans and peaches. It left us with nary a utensil to cook anything in.

The easiest time to extinguish a fire is from midnight until 10:00 A.M. because less wind usually occurs then, the temperature is lower, and the relative humidity (moisture in the air) is higher. After 10:00 A.M. the sun begins to warm things, the dew dries, the breeze springs up, and the fire shows new life. This is why night travel is essential.

How does one man put out a forest fire? First he sizes it up. Are there any snags afire that will scatter sparks? Where is the hot part that needs attention first? Are there bad fuels, snags, slash, or windfalls close by that could be kept from burning by a little work on that side? Often he can cool it down or stop the spread with a few shovels of moist earth. Building a fire line is simply separating the burning materials from the unburned. Anything with live fire is thrown or shoveled inside. Unignited material is thrown outside the line. The fire line is mistakenly called the trench although it is necessary to clean out the line to mineral soil and if there is deep duff the line resembles a trench. Any

Courtesy U.S. Forest Service
LOOKOUT HOUSE ON A WOODEN TOWER
Lindross Hill on the Coeur d'Alene National Forest. The horizontal projections on all sides furnish shade in summer and are dropped and fastened in winter.

burning snags must be felled. After the spread is stopped and you've taken a five, it's time to begin mop-up. That means literally putting it out. There are two good but contradictory rules: 1. Never bury fire. 2. Use plenty of soil to help put it out. Soil chokes the flame and cools the material. It cools the coals under the stick and turning over a burning chunk onto fresh earth slows the burning. Stir earth into the coals. Scrape the fire from burning stumps or logs or chunks with edge of shovel. The boys say, "Use the old smokechaser rub." Handsful of earth rubbed right onto the hot spots on a chunk of fuel is the best possible way to put out the fire. The only safe fire is a dead fire, totally out. It's the unforgivable sin for a fireman to report a fire out and have it come to life again some hot windy day. Sometimes water is real handy and should be used. But water is not a necessity and a good dirt mop-up is just as effective.

Weather forecasting is helpful. Weather, of course, is all-important. Lightning strikes trees and starts fires. Fires start easily after a dry period. In such cases it is hard to put them out. So weather is responsible for dryness of fuels; dryness of air; temperature; velocity and direction of wind. Weather forecasting has improved, the Weather Bureau even has portable units to provide data close to going fires. Relative humidity is used much on the West Coast as an indication of danger. Anyone can determine it by the use of a psychrometer costing only a few dollars. Logging operations are shut down when the air gets drier than 30 percent.

In the Missoula Region Harry Gisborne developed the fire weather stick. It consists of half a dozen round sticks of clear pine 3/8 inch in diameter by 12 inches long and mounted in a frame or holder, like a miniature gate or grate. The oven-dry weight of the sticks is used as a base. The set is placed on a wire rack in the woods about six inches from the ground. It is weighed at regular intervals and the weights recorded. A table gives the percentage of moisture in the sticks at time of weighing. This arrangement gives the composite effect of weather factors and is a true indicator of burning conditions. Weight of the sticks is affected, plus or minus, by rain, dew, sunshine, wind, air moisture, temperature and even the earth beneath.

Some rules that apply to fires, large or small, are:

1. Every fire looks worse from a distance than when you reach it.
2. Fires travel uphill much faster than downhill.
3. Winds blow down canyon in the evening and up canyon in the morning.
4. The sooner a fire is controlled, the easier and cheaper the control is.
5. Fires die down after sundown and liven up after sunup.
6. Every fire has its ups and downs. Thing to do is to subdue it when it is down.
7. The distance around the fire rather than the acreage determines the size of the fire job.
8. Any fire can be put out by sufficient effort and by taking advantage of the breaks.

I realize that exceptions prove the rule. Example: Rule No. 3.

We had about a hundred-acre fire on the St. Regis District in western Montana. It was down the Missoula River Canyon below St. Regis, where the river makes a sharp bend, but on a side stream coming in from the west. A strong wind blew down the river at sundown but was partially blocked at the big bend and the pressure behind forced the air up our side stream. For three successive days we had a line around the fire only to lose it again due to that uphill evening wind.

Group Training

Early it became our regular practice to hold three-day guard training camps. The lookouts and firemen were grouped just before going to their stations. This included intensive training in map reading. The use of the firefinder and the location of fires was explained and practiced. Actual smokes were used. Practice was conducted in the use of compass, and measured distances were paced. Everyone engaged in the "smoke chasing" problem. Three or four smudges were started at points two to four miles distant. One at a time each trainee was dispatched to one of the smokes. Before starting he was given the same information he would have received when going to an actual fire. This might include lookout reading, estimated distance from starting point (or the lookout), estimated size, on which slope— north or south, east or west—and the color of the smoke. He might be advised he could use his compass and go straight across country or he might go part way on a trail, then get back on line when near the fire. A man at the smudge clocked him in. Final training was actually putting out a fire set for training. This included building a fire line, cutting snags, and finally mopping up—putting the fire entirely out. Use of tools, water, and soil was demonstrated and practiced.

At a training camp near the Honeysuckle Ranger Station in Idaho, we had a demonstration of early mop up, instead of the older practice of letting a fire die out. The fire area was not large, about thirty by eighty feet, but it was loaded with dry treetops, limbs, and poles up to twelve inches in diameter. It made a hot fire and took a good small-crew effort to get it stopped without getting too big. When it was proposed that we now go in and put it out—Ranger George Haynes, a good fire man and a good ranger, was dumbfounded. It was hot. Logs were burning. There were deep beds of coals, and there still was much blaze. But we went at it—first shoveling soil, knocking down the flame, cooling off the burning logs. Gradually we worked in from the edges, turning over chunks, providing fresh earth to stand on, stirring soil into the coals, scraping fire from logs with shovel. Soon the fire looked less formidable and in less than two hours it was completely dead. No boots were burned and no one scorched his eyebrows. Had it not been mopped up, it would easily have taken a full day to burn out. The object is to kill it while you have it down, and avoid the possibility of a wind carrying fire across the line.

The North Idaho-Montana Fires of 1910

The northern Idaho fires of August, 1910, were head-lined and publicized nationwide. As a first-year forestry student I read everything about them I could put my hands on. It made a deep impression. Later, in south Idaho, I had a closer contact. George Crockett and Bert Mahoney, rangers on the Minidoka Forest, had been on the fire and Bert told me of their experiences. They didn't arrive until after the big blowup on August 20. At Missoula where they changed trains, they saw thirty coffins loaded on their train

Courtesy U.S. Forest Service
LOOKOUT HOUSE ON A STEEL TOWER
Russell Mountain on the Wallowa-Whitman National Forest. It's a long trip to the woodpile. Many man-and-wife teams take these summer jobs.

headed for Wallace, Idaho. They weren't too sure they were right for a job like that. They had put out a few small fires, but the timber types, the mountains—everything was different. They were cowboy rangers. Like ex-Congressman Sam Coon—Second District of Oregon. Sam's wife, Opal, used to say that Sam was just the handiest man around the ranch. He could do anything—so long as he could do it on horseback.

In 1923 I was made supervisor of the Lolo Forest, and in 1931, I was transferred over the state line from Montana to Idaho in charge of the Coeur d'Alene Forest. These two forests and the St. Joe Forest suffered the heaviest during those terrifying days in 1910. Seventy-two men perished on August 20 on the Coeur d'Alene alone. For eighteen years I lived in the shadow of those billowing smoke clouds. I've flown over the snag fields, traveled the roads, walked or ridden the trails, helped plant barren areas, and relived the holocaust with men who survived it.

Frank D. Haun, ranger on the Savenac District during my entire period on the Lolo, was there in 1910. He was credited with saving the town of Saltese, using a crew of Northern Pacific railroad workers. But he didn't talk much about the big fire. A lot of his district wasn't burned, of course, and very few acres burned or reburned subsequently. He couldn't have been more interested and concerned for that district if it had belonged to him personally. In his crew he had Alex Donnally (who lost one eye) and a couple of other old-timers who had gone through the fire. They were loyalty personified when it came to doing anything for Frank. He started a tree nursery at his headquarters and little trees grown here were planted on some of the clean burned areas. On one visit to his district he and I rode up a trail through the big burn to the Idaho line, along it some distance, and started down another trail, but the trail maintenance crew had missed this one. We had axes on our saddles and started cutting out the windfalls as we came to them. As the time slipped by we began to leave more and more, letting our horses first step over them, and then jump over uncut logs and poles. Finally Frank said, "If we can't throw the bridle reins over them, we'll cut them out." But we had to do a little better than that. The problem was that the windfalls on either side of the trail were so numerous and piled so like jack straws that it was impossible to get a horse through, or over, or around them. Not too long after dark we hit the place where the trail crew had stopped, and we put away our axes.

Elers Koch, supervisor of the Lolo in 1910, was Assistant Regional Forester in charge of Timber Management in 1923. He was an expert forester and a firm believer in the superiority of Idaho white pine. He was much concerned by the take-over of the burn by lodgepole reproduction. The lodgepole pine cones withstand heat and fire much better than those of other tree species. They will hang on the tree limbs unopened for years and years, but when a fire occurs they are opened by the heat and the winged seeds are carried long distances. The higher elevations were supporting lodgepole growth before the fire. A sprinkling of young growth white pine came in if seed trees on unburned areas were near, or where burning of the duff and soil hadn't been too severe.

Koch didn't believe in building up his overhead office at the expense of manpower on the forests. He operated his important division with one assistant, one planting specialist, and one logging engineer. And he found time to visit most of the important sale areas, planting jobs, and timber surveys.

Ranger Ed C. Pulaski was the District Ranger at Wallace, Idaho, Coeur d'Alene National Forest, at the time of the 1910 fire. I met him while he was still active. His health was not good and his eyes, injured in the fire, were giving him trouble and he had retired prior to my assignment to the Coeur d'Alene Forest. Only a limited number of prominent foresters have been so widely known by name as Ranger Pulaski. This is partly due to the axe-mattock fire tool that he devised, the widely accepted Pulaski tool. The name itself is well known. He was a descendent of a long line of

Polish patriots. He was the outstanding hero of the 1910 fire.

Ranger Pulaski, with one hundred and fifty fire fighters, was on the divide south of Wallace in an attempt to hold the fire, which was burning north up from the St. Joe side. When the fire blew up, part of his men were cut off and in danger of being surrounded. He got them together and started to lead them down to Wallace. Another terrific run of fire cut them off in that direction. The only salvation seemed to be to get them into the unused tunnel of the War Eagle mine. They made a run for it. All got to shelter but one man, who perished in the oncoming flame. Water was in the tunnel and the men wet down their clothing and hung wet blankets over the mouth of the tunnel. So much smoke and heat penetrated the one-hundred-foot tunnel that the forty men could scarcely breath and some were about to run out into the consuming walls of flame outside. Pulaski, pointing his revolver in their direction, ordered them to stay inside and lie down on the floor of the tunnel. His own position closest to the mouth was the most critical and in a short time he passed out as a result of the terrible heat and smoke. Most of the men lost consciousness, too. After hours had passed and the flames had subsided, one man was able to leave the tunnel and somehow make his way over the fallen timbers, across burning logs and around beds of hot coals to Wallace. Following his report a rescue party headed for War Eagle. Upon arrival they moved the men into the comparatively cool air outside. Five of the men were dead of suffocation. The others were in bad shape, eyes blistered, hair singed, and lungs affected, but they survived.

The others of Pulaski's crew were able to make it to a high rocky point, far enough from burning timber to escape the flame and heat, but they suffered from the dense smoke that enveloped them.

In the early thirties we built a tower with glass-ribbed lookout house on top, on a point overlooking the War Eagle mine. This was about central in that portion of the big fire area and served to give better detection coverage and shorter travel time to fires. Young tree growth by then had pretty well obscured the down stuff left after the fire. The lookout was named Pulaski Peak.

It is hard to imagine or describe what people went through on that August 20, the day the fire exploded and ran wild on a forty-mile front. Two trains evacuated hundreds of women and children from Wallace over Lookout Pass to Missoula, picking up more folks on the Montana side at Saltese, DeBorgia, and Haugan. The Northern Pacific had no rails west of Wallace. The Milwaukee Railroad evacuated people from Avery west to Spokane. The railroad bridges and ties from Avery east to the Taft tunnel and for twenty miles on into Montana were on fire. The fire consumed 30

Photo by Carl Holman Studio, Baker, Ore.
A FIREMAN'S PACK
The contents are displayed on the board. Included are three nonperishable meals. Below, the pack is assembled, ready for a quick getaway.

percent of the town of Wallace, the county seat and largest town in the county. The villages of DeBorgia, Haugan, and Taft on the Montana side were wiped out. Prospectors, miners, fire fighters, and others were trapped. The fate of many was not known for days. Some found ways to save themselves but others died.

Supervisor Weigle, whom I knew later, was out from Wallace on horseback before the blowup, checking on conditions and helping residents. He had requested the railroad to hold a train in Wallace should it be needed. When the blowup came he was cut off. His saddle horse died in the fire and he first sought safety in a tunnel, but the timbers caught fire and he was forced to get out. He partially buried himself in sand and earth and managed to survive although burned about the head and back. He later went to Juneau, Alaska, as Supervisor of the vast Alaska forests.

Probably 80 percent of the Coeur d'Alene Forest has quite gentle slopes, is well soiled, and is unique in that timber grows in good productive soil even right over the main ridge tops. It has very few rocky mountaintops. The other 20 percent—coinciding quite closely with that area burned in August, 1910—lies in the extreme southeast corner of the forest up against the Bitterroot Range, which there marks the state boundary line. It is terrifically rough—not so rocky that it doesn't produce rather heavy and dense stands of conifer timber—but extremely steep with narrow gulches between the hills. How the soil stays in place is a marvel. It surely wouldn't with rainfall comparable to that on the coast or in the rain forests. A good illustration is the mining town of Burke. The canyon is so narrow and the slopes so steep that the railroad goes right through and under the ore mill, which is built partly on each hillside. And the railroad and highway are compelled to share the same roadbed for some distance.

This kind of topography provides the toughest kind of fire fighting imaginable. But, perhaps, on August 20, it didn't make much difference. When the fires roared up one hill there was so much wind and swirling draft from the fire itself and walls of flame and burning gas and wind-borne material in the air, that it carried right on down into the next gulch and up the next hill. If any area on the lee side escaped temporarily a reverse current soon consumed it from below.

Bald Mountain Fire (Selway-1929)

This fire started on the Lochsa District of the Selway Forest about one hundred and twenty miles east of Lewiston, Idaho. It was in a big, roadless, almost wilderness area accessible only by trail and miles from supply points in any direction. This was in August, following a hot dry period with numerous electric strikes. As many as one hundred lightning fires have occurred in one day on a single ranger district there. This particular fire escaped the initial suppression force, blew up and spread alarmingly. This called for large-scale reorganization and, due to transportation difficulties, the Selway Forest took the west end and southwest side, the Clearwater Forest was assigned the north side, and the Lolo Forest fell heir to the east end and southeast side. That was the front end, uphill, and with the wind against us. Two ranger districts, the Powell and Elk Summit at the east or upper part of the Selway Forest had been turned over to the Lolo Forest for administration from the Montana side because they could be more readily reached from Missoula by way of Lolo Hot Springs and Lolo Pass. The two units were then spoken of as the East Selway.

On the day that the Bald Mountain fire blew up I came into Missoula from a week on fires down the river on the Lolo, spent an hour in the office and headed over Lolo Pass for the Powell Ranger Station on the East Selway. It was evening and I'd been on short rations for sleep. As I rolled along the one-lane road down the Crooked Fork, I caught myself nodding, but I continued. A mile farther on I went soundly to sleep at the wheel but only for a moment. A quick application of the brakes avoided a bad spill into the stream below. A vigorous run up and down the road, some jumping up and down and some arm beating got the blood circulating and the cobwebs out.

At the Powell Ranger Station, the end of the road, I was joined by Ranger Ed Mackay and we started on foot down the Lochsa River trail for the fire. As we approached Colgate Licks, we met a fellow walking toward us wringing his hands and crying, "Help! Help!" adding as we neared him, "I've rolled my mules and I've lost my saddle horse." He turned around and went with us a couple of hundred yards and, sure enough, there were nine head of big mules piled up in the gulch. All were heavily loaded with fire equipment, cooking outfits, groceries, and fire beds. Our flashlights disclosed legs, feet, heads, packs, ropes, tails, all in a tangled heap. One or two struggled a bit, but most were motionless. Now and then one that was hurting would make a plaintive cry, more like a human than an animal. We went to work, Ranger Mackay doing the most. One at a time the top mule or the one easiest to get at was released by cutting a lead rope, a cargo rope or sometimes a latigo, and he was helped to his feet and led up on the trail out of the way. This continued until every mule was back on his feet. There wasn't a broken bone or a crippled mule. Some of the rigging was badly disfigured and the beans and bacon and things were in disarray. But the packer was able to put together better than half of his freight. We found his saddle horse a short distance down the trail and left him to reload—his confidence restored. Ed was a real hand, not only with mules but with men.

The trail was a laid-out trail—that is, it was on grade. When it came to a draw or swale it made a curve around on the level instead of a dip down into the low ground and out again. The tendency in the latter situation is for the mules to speed up going into the dip. Those following fail to speed up quickly enough, which causes broken ropes and trouble. Usually a "pig tail" or light rope loop is put on each pack saddle and the lead rope is tied to the pig tail ahead. This saves the lead rope and the saddle. In this case the mules failed to follow around the curve at a uniform speed and the caboose pulled the whole train off the track. The pig tails didn't break as they were supposed to.

A few miles farther on we met another man on the trail. In a half scream he said, "It's a-coming." One of us asked, "What's a-coming?" "The fire," he shouted, and added, "It'll never catch me. I'll cut my throat first, and I've got a razor in my pocket to do it with." He was out of breath, his eyes were bugged out, and he was really in a panic. We tried to reassure him and told him we were going on down.

A half mile farther and we met a fifty-man crew. Chet Olson was in charge. He, with other forest officers from the Ogden Region, had been detailed to Missoula to help on fires in our region and he had gone down the Lochsa with the initial crew that day. In later years Chet was to become Regional Forester of the Intermountain Region. One of the men was draped in a new U.S. flag salvaged from the fire

Courtesy U.S. Forest Service
RANGER FRANK HAUN, VETERAN OF THE FOREST SERVICE AND HERO OF THE DISASTROUS IDAHO-MONTANA FIRE OF 1910

kit. Each was carrying a fire bed or a tool. They had put most of their fifty-man outfit in the river to save it.

The fire had been moving upstream and it looked too dangerous to them to make camp. Besides, the fire had jumped the river and was advancing on the south side also. We talked over the situation with Olson. It was about 2:00 A.M. and we decided to stay right there until daylight. The river was low and we found a rocky untimbered beach. Those with fire beds spread them out and lay down on them. Those without found the softest ground they could and stretched out. No one got into bed. The fire to the southwest cast a fluttering light on our beach camp. Now and then a snag would crash, or fire would crown out in a patch of young saplings. At these noises a half dozen or more men would sit upright and stare up at the hillside. They resembled a bunch of jacks-in-the-box. No one got much sleep. But the night wore on without more excitement and with the coming of daylight, they went down the trail with us.

A base camp was set up at Jerry Johnson Bar.* A telephone was there and some open level ground, and at that point a trail went out each way from the canyon bottom.

More fire fighters were sent in to us. Several remount pack

---

* Jerry Johnson was an historic point. Lewis and Clark, on their trip west in 1805, had stopped there. Their route from Missoula and the Bitterroot River led to Lolo Hot Springs and Lolo Pass. This put them on top of the Bitterroot Range, later to be the boundary between Montana and Idaho. They worked their way down a timbered canyon to a point known now as Powell Ranger Station. Here they killed a colt for food and named the stream nearby "Colt Killed Creek." It is now shown on the maps as White Sand Creek. They found no game here, while now deer and elk are plentiful and moose are found, too. They continued on down what is known as the Lochsa River to Jerry Johnson. Their scouts reported the river route beyond was too tough and the party climbed out of the canyon in a northwesterly direction to the main east and west divide at Indian Post Office and followed the ridge west. The Post Office was a prominent point bearing some rock cairns, reportedly used as a smoke signal station and marker on the Lolo Trail where the Indians left the trail to go to fishing grounds on the Lochsa River.

Later, in the fall of 1893 the Carlin hunting party camped here and hunted elk—killing many. They were snowed in and couldn't get out with their horses. Then they built two rafts and attempted to go downstream. But they were wrecked and their supplies were lost. They left their cook, who was ill, at the big cedar grove, where he perished, alone. Finally, after a gruelling trip afoot, exhausted and almost starved, they were rescued by a detachment of soldiers from Fort George Wright, Spokane.

strings were trucked to Powell and packed outfits and supplies to the camps. They also moved some camps up out of
the canyon. Additional overhead was sent in from the Regional Forester's office and other forests. Bulldozers and
power saws and pumpers were not invented then. We had
to do it the hard way. On the flats were big cedars, some six
feet through. On the north slopes, were fir and tamarack,
nearly three feet on the stump. On the south slopes were
pine and fir. We had no walkie-talkies. Scouting had to be
entirely on foot. Bone-weary, sleepy scouts and fire bosses
held night huddles and decided on tactics for the morrow.
We lost lines. We backed up and started over. We abandoned camps. At one time we had seven, twenty-five-man
camp outfits in the Lochsa River to keep them from burning. We built lines close to the fire. We backed off and built
faster, straighter lines and burned out in between, but we'd
still lose out. There always seemed to be too much fire or
too much wind or too many burning snags. On the seventh
day we built a good line, up the Jerry Johnson-Indian Post
Office trail, and had men and torches and fuel oil scattered
out to start a couple of miles of backfire. We waited for
the fire to advance. The fire itself was half a mile away. With
its usual perversity but with different tactics this time, the
fire didn't advance. There was more moisture in the air and
the wind wasn't so severe. So we were able to jump in with
all the drive we had left, and shove a line right up the fire's
edge and hold it. The good backfire base was never used.
And we held the line on the south side of the river, too.

We lost one man. But he wasn't burned to death. His
skull was crushed when a chunk fell from a burning snag.
A couple of days after his body had been sent out to town,
someone found his false teeth where he had fallen.

Sometimes a lighter vein offset the more somber events.
Tom Lomasson had a crew and camp high up on the south
side of the river. One day the packer brought down a note
from Tom. All he said was, "Do all of those Missoula steers
walk only on front legs?" We got our meat by the quarter

and some camps closer in, had evidently been getting more than their share of the hind quarters with the rounds and loins.

A blacktop highway has now been completed along the Lochsa, making a through connection between Lewiston, Idaho, and Missoula, Montana. Ranger Mackay, now retired, writes that it is a beautiful drive and a wonderful road—at least he thinks so compared to covering it at four miles an hour with a saddle horse and a string of mules. He says the Bald Mountain fire left quite a scar and, farther west, the country was badly burned in the 1910 fires.

Along this recently completed Lewis and Clark Highway is a roadside park marked in memory of the conservation writer, Bernard DeVoto. It is located on the Crooked Fork of the Lochsa River a few miles northeasterly from the Powell Ranger Station. Here DeVoto camped a number of times while retracing the steps of Lewis and Clark. He became so enthusiastic about the country and its historic interest that he arranged before his death to have his ashes scattered over this locality.

In early 1965 the *Oregonian* reported passage by the Senate of a bill authorizing the Nez Perce National Historical Park. But instead of carving out a single block, it would create twenty-two little islands of historic interest to commemorate the great Nez Perce Indian Wars with the Army in 1877, and the Lewis and Clark expedition through Idaho in 1805. The sites were not reported but certainly the Powell Camp at the mouth of Colt Killed Creek and Jerry Johnson Bar (gravel bar, that is) would be two of them.

*The Tillamook Fire—1933

This fire is not properly a part of this narrative as our only contact with it consisted in smelling and trying to penetrate the smoke at a distance of more than four hundred

---

* Stewart Holbrook, *Burning an Empire; the Story of American Forest Fires; with a* foreword by William B. Greeley (New York: The Macmillan Company, 1943).

miles. It was tremendously important, though. While not
the greatest in acreage, it undoubtedly killed or consumed
more timber than any fire in known United States history,
as it was in big, old-growth Douglas fir timber in the rain
belt of the Oregon coast. Still, its size was extreme, too. A

Courtesy U.S. Forest Service
SLEEPING CHILD FIRE, MONTANA, 1961

township of land is six miles square and contains 23,040 acres. The Tillamook fire covered 13½ townships. That is nearly a third of a million acres. And 87 percent of the acreage was burned over during one twenty-four-hour period on August 24, and 25, 1933. Yet only one life was reported lost.

It is reported that as much timber was salvaged from the Tillamook burn as was estimated at first as killed. This was largely due to rising log prices. It was further assisted by big new markets, such as pulp, particle board, and pressed wood. Salvage material was readily usable. Logging fire-killed timber is dirty work. At night the loggers were black as any negroes. But fire-damaged timber is not the total loss it was formerly.

This is the last huge fire to date. May that statement hold good always!

Other Smaller Fires

The Nine-Mile fire, 1926, on the Frenchtown District in western Montana, was a preseason fire on private land, threatening government timber. This case went to court, the judge requiring the landowner to pay the costs.

The Quartz Creek fire, 1926, on the Quartz District, Lolo National Forest. While this was less than a hundred-acre fire when it was stopped, it made the loudest explosion and roar I ever heard from a fire. Ranger George Hankinson and I were right beside it when the fire hit an extremely steep slope covered with dense young growth and fine fuels. George was slightly hard of hearing but no repeat was needed that time.

The Low-line fire, 1928, on the St. Regis District. This was the one where the down-canyon evening wind made the error of pushing uphill on a side stream. The first evening was a surprise, the second time was a mistake, and the third time it became a habit.

The Trout Creek fire, 1930—Superior District, western Montana. A U.S. Congressional delegation observed this

fire from a safe distance while traveling by special bus on
U.S. Highway 10. I left the bus and junket at Superior, in
my town clothes, bought some work clothes, and went to
fighting fire. My main recollection is that I simply wore out
a new pair of ordinary work shoes before I got home to my
logger boots.

The McPherson fire, 1931—Coeur d'Alene Forest. This
was largely a reburn of an area burned in 1910. On the
Pacific coast they dread an east wind because it brings dry
air from the interior. In our country bad winds, and in fact
most of our winds, blow from the southwest. A southwest
wind took this fire out of our hands and presented part of it
to our neighbors to the northeast on the Cabinet Forest in
Montana.

The Big Cow fire, 1939—Blue Mountain and Unity Districts, Whitman Forest, Oregon. This, too, was an interforest

Courtesy U.S. Forest Service
OLD AND NEW MEANS OF SUPPLYING FIRE CREWS—ALL IN ONE PICTURE

fire. It started on the Malheur Forest to the southwest. It ran up to the summit and refused to recognize the forest boundary, coming three or four miles down on the Whitman side. On our side the fire was plenty hot enough to kill most of the timber but not to consume it. One of the compensating factors was the success of the salvage operation. About 95 percent of the timber was utilized. One man on Bald Mountain lookout east of Prairie City lost his car because of this fire. He stayed in the lookout reporting its progress and then left to go down to his car. When he arrived there, fire was all around and he ran back up the steep hill to the lookout, injuring his lungs. The car was destroyed.

After this fire, G. D. Pickford, U.S.F.S., and E. R. Jackman, then with O.S.U., made numerous grass plantings that helped to form the after-fire policy in this region.*

Jim Hutchens, fire control officer on the Glacier View District of the Flathead National Forest, in northwest Montana contributes the following:

A Fire in the Bob Marshall Wilderness

Late in September one year when the weather had been exceptionally dry, fire was reported in the Bob Marshall Wilderness Area. We had to be flown part way and walked the last 26 miles.

It was hard to get fire fighters so we recruited around 100 Blackfeet Indians and they walked in to the fire. Everything went well except for the many Indians who had worn their shoes out on the way in. This was taken care of by ordering new ones for them. On the fifth day we had the fire well under control and were talking of letting some of the Indians go home. The next morning when we woke up there was a good six inches of snow on top of our sleeping bags and more coming. It didn't take long to decide that

---

* Their joint bulletin, published in 1944, was: *Reseeding Eastern Oregon Summer Ranges*. Station Circular 159, Oregon State University.

everyone, except a few, could go home. The nearest way out
for the Indians was over the Continental Divide and down
to the reservation. Four Forest Service men went with them
to be sure they all got out. When they got to the top of the
divide it was snowing and blowing real hard with the temperature around zero. It was hard to see more than a few
feet ahead due to the driving snow, as they were facing into
the wind. A couple of Indians were quite old, and they
would get off the trail every chance they had and get under
a tree. When the rear guard would find them they would
say they were going to die right there. The Forest Service
men would get them going again but every time they would
let them out of sight the old boys were in the brush again.
They did get out with no casualties.

The other Forest Service men had gone out another way
leaving eight of us to pick up tools, pack grub, and help
get the food and tools for 150 men out to the station where
the plane could pick it up. Among this food that was left
because of the sudden change in the weather were several
hind quarters of beef and among the eight men were two of
the best fire cooks that have been found in many years of
fire fighting. Every morning when they would call the
rest of us to breakfast there would be fine large beefsteaks;
when lunch was served there were more beefsteaks, and when
the evening meal was ready there it was—more beefsteak.
After six days of this kind of torture the day came when
everything was packed out but the gear we were using, so
the cooks fixed a nice lunch of beefsteak for us to take along
and we all walked that 26 miles out. When we got to the
station that evening we found the cook there had something
ready for us to eat. We sat down to the table and found
a pot of what he said was beans. We found a very few beans
and one small piece of ham floating around in the soup. We
knew we had sent along some quarters of beef but found
it hanging up in the meat house; the cook said he was saving
it for when they needed it.

For Fewer and Smaller Fires

The great big majority of all the fires that start in the forests are put out as wee small ones. Each national forest in these western mountain regions handles annually from seventy-five to two hundred and fifty fires. Nearly all are put out as Class A fires. A Class A fire is less than a quarter acre in size. Many are kept within that size through faithful, conscientious attention on the part of lookouts and some strenuously grueling hikes and fast work on the part of firemen. With that kind of spirit and modern improvements, it is possible to hold the percentage of burned forest land to a low figure. In our day we were handling fires on a horse and buggy basis. Use of airplanes and helicopters has improved detection. Smoke jumpers speed up first attack. Parachute delivery of tools, equipment, and supplies is faster than pack mules. D-8 caterpillars, dozers, and power saws facilitate line construction and falling of snags. Radios speed up communication. Walkie-talkies allow fast communication between scouts and fire bosses and crews.

For years the goal has been to have every fire under good control by 10:00 A.M. of the first day. We didn't always get it done. But the goal is closer every year.

More babies are born every year. That means more visitors to the national forests. More cars, more miles of road, more leisure time, all add more travelers through the forests. Chances for man-caused fires are greater. That means more need for fire prevention work. Airplanes and bulldozers have advanced the suppression work but something else must be used to reduce man-caused fires.

One teenager from among five hundred may be the only law violator but he gets bigger headlines and more publicity than all of the four hundred and ninety-nine combined. So it is with fires. One Class C fire gets photographed and publicized in a big way. Nothing is heard of the ninety-nine Class A's. Perhaps a third of them have an interesting story. Detection and/or suppression by a volunteer, maybe a Boy Scout or a prospector. Maybe a fireman has made an unusually long or tough trek, or by real skill has held a fire about to get away. Perhaps the starting of the fire illustrates how not to use fire on a camping trip. No two circumstances are the same.

The drama is in the big soul-searing episode, whereas the little fire, well handled, draws no headlines. Many of the small fires carry the best lessons, illustrating proper procedure, cooperation between the public and the Forest Service, and intelligent use of the tools at hand.

Courtesy U.S. Forest Service
SMOKE JUMPERS LOADING INTO A DC-3 PLANE
FOR A QUICK TRIP TO AN ISOLATED FIRE

CHAPTER FIVE

BLOCKING UP

CONGRESS, IN 1905, CREATED THE NATIONAL FORESTS AND designated the United States Forest Service to handle these far-flung lands. Not counting Alaska, they totaled roughly 180,000,000 acres. Until that time the whole tendency was to get public lands into private hands. Railroad land grants, military road grants, timber and stone claims, scrip purchases, Veteran grants, mining claims, several kinds of homesteads, school grants, gifts of many kinds—all of these contributed to the passage of millions of acres into private ownership.

This was a natural process that assisted in settlement and development of virgin territory. The major appropriation of land occurred in the fertile valleys and adjacent foothills and choicest, most accessible timber tracts. When the forest boundaries were drawn, as much of the deeded ground as possible was left on the outside. Still every forest included hundreds of tracts of privately owned land intermingled with the government ground.

The administrators of these young Federal properties found their work complicated by the presence of these scattered private tracts. Sale of timber, handling of domestic stock on the range, protection from fire, road building, fencing—all of these things were done less efficiently and some needed jobs could not be done because of the existence of private land. Forest maps were printed to show government land in light green and private land in white. Every white tract made potential problems.

In recognition of these situations and to further assure the conservative management of timber-producing lands, Congress passed the Act of March 22, 1922. This authorized acquisition of suitable private lands within national forest boundaries in exchange for either government land or government timber. No funds were provided for outright purchase.

How It Worked

Various situations developed. The most frequent type of applicant was the operating lumber company. It owned cutover lands that it wished to dispose of and it was purchasing national forest stumpage. Cutover lands offered

Photo by Bob Bailey, Enterprise, Ore.
TYPICAL MATURE PONDEROSA (YELLOW) PINE TIMBER
The county lost $25.85 per year in taxes on this quarter section turned in to the government. It stands to gain about $15,000 through the 25 percent fund when the timber is cut.

were examined, mapped, and a descriptive report prepared. Values were established. Exact procedures were worked out with the lumber company.

Each exchange case was advertised for thirty days in a paper with general circulation in the county in which the land or timber was located. If in two counties, it had to be advertised in both counties, to give public notice of pending exchange and to provide opportunity for protest. Proof of publication was filed with the records.

Shortly after 1922 it was a requirement to get written approval from the county commissioners in each case before the final transaction. I have presented cases to seven county boards in three states, Montana, Idaho, and Oregon, and concurrence was secured in every instance. An occasional commissioner said that he disliked to see the county lose taxes, but he thought the exchange was for the best in the long run.

In other instances land was exchanged for land if it was clearly in the public interest. It was the practice to dispose only of those parcels isolated from other government land, along the forest boundary, or land with low value for forest purposes. It is often said in the West that any rancher can see how it is to his advantage to "round out his holdings" and that all he wants is the "land adjoining him." This isn't the case with national forests.

The majority of small timberland owners simply wanted to dispose of their land. This resulted in a third situation, a three-party or tripartite exchange. The landowner turned over his tract to the government for certain timber. That timber the Forest Service then marketed for him, collected the stumpage payment and passed it on to the former landowner. These cases almost always involved only small units of about 160 acres.

In one such instance, Dr. E. B. Young, of Baker, Oregon, turned in 160 acres for $1.50 an acre. The cutting of selected timber had been delayed and the payment to the land exchanger had not been made. In a conversation about it with the doctor (a well-to-do man) in the presence of his two daughters who were graduating from high school, he remarked, "That's too bad. I'd been counting on that to send these girls to college this fall. If it doesn't come soon I guess the girls can't go to college this year."

In another instance Fred T. Sterling, president of the Western Montana Bank of Missoula, Montana, had made application to turn over about nine hundred acres of land within the Lolo Forest. It was in scattered small tracts, some in the 1910 burn and some was tax-title land or land acquired through foreclosures. He had agreed upon a valuation of seventy-five cents an acre. When I took up the title problems with Mr. Sterling he threw up his hands and said, "Forget it. I'll let the land go for taxes. I won't get much for it anyway." I offered to see if I could clear up the titles if he would pay any necessary expense. He agreed to spend up to fifty dollars. The land was in Mineral County, with the county seat at Superior. Former owners had made timber reservations when they sold their tracts and this was one

Courtesy Div. Blister Rust Control, Bur. Plant Industry, U.S.D.A.
IMMATURE OR POLE STANDS NOT TOUCHED DURING LOGGING
These sections were often acquired along with logged areas at $1.50 to $2.50 per acre

of the principal clouds. These reservations in every case had had a time limit, usually ten years, and that limit had long since passed by, thus nullifying the reservations. Another matter was cleared by a statement that Fred T. Sterling, Fred Sterling, F. T. Sterling, F. Sterling, Frederick Sterling and Frederick T. Sterling were one and the same. None of the fifty-dollar budget was expended.

The Whitman National Forest in eastern Oregon increased its net acreage of government land through land and timber exchanges by 235,731 acres, very largely during the period from 1938 to 1952. Seventy-two cases were handled. This is equivalent to one average size ranger district. No increase in personnel to administer this additional land took place. During the same period the number of ranger districts was reduced from seven to six with a corresponding decrease in personnel. Part of this saving was due to other causes, but a substantial saving was due to blocking up the ownership. Problems are eliminated every time a white spot on the map can be colored green.

We did not accept applications to turn in patented mining land. If there were enough mineral showings to justify the issuing of a patent in the first place it is likely that it would be taken again by a new locator. The mining laws apply to lands acquired through exchange just as they do to other national forest land.

The same thing is kept in mind in the case of patented homesteads. If they actually had agricultural value we did not acquire them with the risk of losing them later.

Loss of Taxes to the Counties

Recently I heard a statement made in open meeting that over 50 percent of the state of Oregon is owned by the Federal government and no taxes are paid by the United States. This is less than half the story. Half of the acreage is far different from half the value. An appreciable part of the national forest acreage consists of rocky mountain peaks, lava flows, old burns, cutover lands, and other low-dollar-value areas. The Bureau of Land Management (formerly the Grazing Service) holdings were available for the taking until June 28, 1934. They were not taken into private ownership in spite of being available as homesteads (160 acres), desert homesteads (320 acres), or grazing homesteads (640 acres). For over seventy years these lands were offered free and no one wanted them. If by some mystic wand all of the United States government land in Oregon, for example, were to be transferred to private ownership tonight and put on the tax rolls, the increase in tax receipts would be disappointingly small.

None of the city taxes would be assessable against forest land. This would throw out taxes for water systems, sewer systems, fire departments, libraries, city schools, city police,

Courtesy U.S. Forest Service
CLOSE CROPPING ON PRIVATE LAND MAY TEMPT CATTLE TO BREAK
THROUGH FENCES TO BETTER PASTURE OUTSIDE

and the like. In a million acres of national forest land there is not a single schoolhouse. The Federal government builds and maintains the highways and roads in the national forests and national parks.

Forest Receipts in Lieu of Taxes

When the law was passed authorizing the establishment of national forests, Congress wisely provided a substitute for taxes. It specified that twenty-five percent of the gross

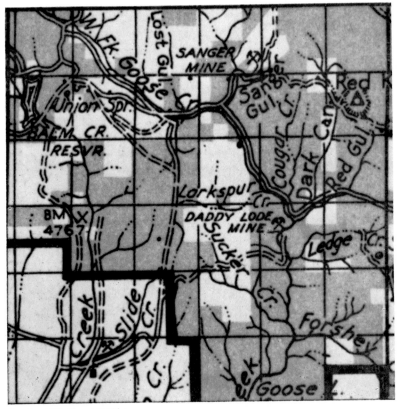

Courtesy U.S. Forest Service
A SOMEWHAT TYPICAL TOWNSHIP IN THE COMMERCIAL TIMBER AREA
OF THE WHITMAN NATIONAL FOREST, OREGON
The shaded area is government land, the white area, private land (1941)

receipts from all sources on the national forests must be turned over to the counties for schools and roads. The government does not determine the allocation between these two purposes. The distribution between counties is arrived at as follows: 25 percent of the annual receipts from a single forest are pro-rated to the counties within which it is located on the basis of national forest acres within each county. Twenty-five percent of all receipts from newly acquired land goes to the counties too.

The law specifies that 10 percent of all forest receipts are

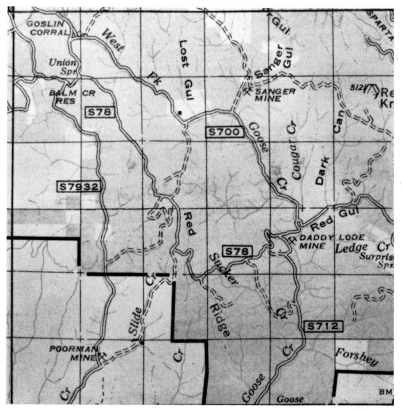

THE SAME TOWNSHIP NINE YEARS LATER, SHOWING OWNERSHIP
AS OF 1952
White areas are privately owned. Between these two dates, the government was actively acquiring isolated tracts of privately owned lands to reduce administrative costs.

to be returned to the Forest Service for use on minor roads and trails within the forests.

In earlier years when much of the logging was on private lands and stumpage prices were low, the 25 percent receipts to counties were correspondingly low. They have increased substantially as these two factors have changed. In fiscal year 1964 the counties in Oregon received $13,-645,118 as their share of national forest receipts.

National Forest Expenditures

Even more significant than the receipts fund in the benefits to the economy of states like Oregon, with large national forest holdings, are the very sizable Federal expenditures in the protection, management, and development of the forests themselves. Of course the county treasurer misses the pleasure of writing these checks. This money comes not alone from Oregon and other Western mountain states but largely from the big successful corporations nationwide and the businesses and people in Chicago, Pittsburgh, New York City, and other centers of population.

There are fourteen National Forests in Oregon and six in Washington, known as the Pacific Northwest Region (Region No. 6).

The expenditures in this region for Fiscal Year 1964 were:

	Operating	Capital Investment
National Forest Protection & Management	$21,139,283	$ 6,343,407
Fighting Forest Fires	2,270,581	8,720
Insect and Disease Control	289,100	4,043
Roads & Trails, Const. & Maint.	7,152,916	14,888,443
Flood Prevention & Watershed Management	49,318	2,359
Coop Work (deposits for stand improvement, etc.)	29,290	3,987,816
Total	$30,930,488	$25,234,786

Acquisition on Whitman in Eastern Oregon

The Stoddard Lumber Company, Baker, Oregon, had been in operation many years. It was owned by a banking family in Utah. They could see the end of the operation. Its life was extended for several years by exchanging its cutover lands for government stumpage. In its final exchange in 1942 the company included 160 acres of virgin timber considered too far from the company mill to be practicable to handle. A valuation of $2.75 per thousand feet of pine was agreed upon. Sherm Feiss, then Timber Management staff man on the Whitman, made a 100 percent cruise and found 1,825,000 feet board measure of ponderosa pine plus some Douglas fir. It was as fine a stand of mature yellow pine as could be found anywhere. It straddles State Highway 26, a blacktop highway, and is only eight miles from a sawmill. At the peak of lumber prices in the late fifties, the Forest Service could have sold this timber for $35 per M feet on the stump. This would have meant a profit on that one quarter section of $58,857. The counties would have received 25 percent of the total sale price or $15,969. The company paid the county $25.85 taxes on this particular tract the last year it was owned by the company.

The Oregon Lumber Company of Baker, the Burnt River Company of Baker, Mt. Emily Lumber Co. of La Grande, Oregon Trail Lumber Co. of Union, all operating companies, made a practice of exchanging cutover lands a block at a time to the Forest Service for stumpage. "Cutover" in eastern Oregon of any yellow pine and associated types is not synonymous with "clean cut." When many of these exchanges were made, white fir was considered worthless and was left standing. Only the most desirable Douglas fir and tamarack were taken out. Even in the case of yellow pine, the immature bull pine, the forked or crooked trees and those with a low percentage of clear lumber (limby) were not cut.

This is well illustrated in the case of the Collins Ponderosa Lumber Co. It was nearing the end so far as its own timber was concerned and it had been turning in its cutover lands. Toward the last it became apparent that the trees passed up in logging had a potential value. The agreed valuation on some of these lands reached as high as $2.75 an acre, the highest local figure at that time. Within seven years white fir found a ready market, Douglas fir and larch were profitable to handle, even the larger lodgepole could be removed. The company roads were mostly usable, the haul was short, and the ground not so steep as on other government land. The same company began to buy back the timber that it had earlier owned and passed by in its logging. Hundreds of acres brought in as much as $40 an acre for stumpage and the government still had the land and immature timber and young growth. These exchanges turned out to be profitable both to the government and to the counties.

The counties got 25 percent of the $40 an acre received for the stumpage cut from newly acquired land. This far exceeded the taxes lost to the counties plus another loss suffered by the counties (25 percent of the $2.75). It was often overlooked that by using government timber to acquire private land, the cash receipts were reduced. This, of course, meant a reduction in the 25 percent fund paid to the counties.

The Forest Service land exchange program was practically stopped in 1952. The statutes are still in effect and many holes in the doughnut still exist. While land values have gone up, stumpage values have risen, too, so further exchanges are feasible and needed. As a result of this legislation and the transactions carried out under it, the national forests are much better management units than they were. Substantial acreage formerly in widely scattered and not strong ownership, has been put under multiple use and sustained yield management. The national forests are a reality. They have had the support of the people and of Congress for sixty years. The exchange-acquisition program made them more valuable, more useful, and more economical. The counties made some sacrifices initially but already they have been more than compensated.

CHAPTER SIX

TIMBER-R-R OUP THE HILL!

Logging Operations

LOGGING VARIED GREATLY IN DIFFERENT STATES AND POWER changed from oxen to steam and then to gas. The time-honored call of warning for a tree about to fall is "Timber'rr" but in Minnesota the predominant Scandinavian timber faller added "oup the hill" or "down the hill."

Early Minnesota

I spent part of 1911 in the white pine belt of northern Minnesota. Only white pine was logged. Five and one-half foot, stiff, crosscut saws were used to fall and buck the trees. Logs were cut into sixteen-foot lengths with a few fourteens and twelves in top logs or at forks. Horses were used entirely in skidding from the stump to a landing on a sled road. Loading was done with a crosshaul. Two skids (large poles) formed an incline from the ground to the top of the load. A team of horses on the opposite side pulled the logs onto the sled but with the chain rigged to roll the log rather than by moving it by a straight pull. Bunks on the sled, twelve feet long, provided for loads of that width. Logs would be piled nine or ten feet high on the bunks.

The sled runners traveled in iced ruts. They logged only in the winter and the log roads were laid out on downgrades and across swamps to a river landing. Where there were pitches steep enough to cause the loaded sled to push the horses, notches were cut across the ruts at intervals and a wire-wrapped bundle of hay the size of a piece of stovepipe was laid in each notch. This served as a brake. A road "monkey" put in and took out hay as needed. All the hay would be out for uphill travel on the return. Water tanks sprinkled the ruts at night. Four-horse teams were used and to help start a load a heavy wooden maul was needed to jar the runners as the teams leaned into the collars. In most cases logs were decked on the ice ready for floating to the mill when the spring breakup arrived. In some locations delivery would be made to a railroad for rail haul to the mill. Cant hooks were much in use both in loading and unloading.

A key building in any logging camp was the mess hall. The kitchen was at one end and no woman ever set foot in there. Women and whiskey were strictly taboo in camp. Two long tables were loaded with food: fresh bread, beans, potatoes, other vegetables, meat in ample supply, and a profusion of pastries. Pie three times a day, fresh doughnuts, and cookies of all kinds were served. A good cook believes sugar and flour cost far less than meat. No lettuce or grapefruit in those days.

Not a word was spoken during the meal. Etiquette allowed one to reach as far as he could but forbade asking to have anything passed. In an unbelievably short time the men were finished and climbing over the wooden benches to leave. Toothpicks were in a box by the door. A flunky took hot noon meals to those out in the woods too far away to walk to camp. This was no lunch but a regular dinner served hot, around a blazing fire.

Bunkhouses were tar-paper covered and practically windowless. Bunks were known as "muzzle loaders." They were built solid down each side of the building, one end against the wall and two bunks high. Thus to enter your stateroom you crawled over the footboard. Your sitting room was the same end of the bunk with your feet hanging out. Eight-inch boards on sides and ends kept some loose straw between you and the lumber bottom. The straw served as both mattress and springs. Usually three blankets were issued upon arrival. It was undignified to pack your bed from job to job. A large wood-burning heater was just inside the door at each end of the building and a couple of coal-oil lanterns made an attempt at illumination. A building sixty feet by twenty feet housed eighty men. A newcomer looked up, saw a couple of cans on top of the stovepipe to provide more draft, and said, "I've seen haywire outfits and gunnysack outfits, but I never saw a lard-bucket outfit before."

The toilet consisted of a deep pit some thirty feet long with a husky pole supported along the front. A shed roof covered the pole and pit. Flies were no problem in the minus-forty-degree weather, but that pole was a bit frosty at five o'clock in the morning. Users did not loiter. The place

Courtesy Calvin L. De Laittre, Minneapolis, Minn.
WHITE PINE LOGS EN ROUTE TO A LANDING
Frank Lydic, teamster, northern Minnesota, winter 1900-1901. Sleds ran in iced ruts.

provided neither privacy nor suitable atmosphere for meditation.

The horses were well stabled, well fed and well groomed.
They worked hard. I've seen many nosebleeds among the
horses due to pulling below-zero air into their lungs during
heavy exertion.

The office was small and no place to loaf, either. You
went there for your time when you quit. And you got your
snuff (Copenhagen), socks, mittens, pipe tobacco, and such
necessities from the timekeeper who ran the wanigan. Such
purchases were charged against your time.

I've seen some of these men patching their old underwear
and darning their wool socks at night and wondered about
human nature, knowing that when spring finally came and
they got to town they would squander their whole winter's
wages in three or four days of riotous living. One Swede
said when his pockets were empty—"Oh, vell! Easy come,
easy go." But they gave an honest day's work and were proud
of it, glorying in their strength and skill.

For generations loggers lived in isolated camps and communities and created their own distinct manner of speaking.
But their jargon is fast becoming archaic.*

Nevada

Later across the continent I encountered another timber
operation like nothing else I ever saw. Perhaps it was because mules were involved. And over the years mules have
grown on me and have won real respect. In some ways mules
are smarter than humans. This activity was way out in the
middle of Nevada, miles from nowhere. A mining company
had built a mill for crushing ore and concentrating the
mineral. It was a steam power plant and the fuel was piñon
pine cut into lengths a little shorter than cordwood. It was
packed in from the mountains on mules. Packsaddles were

---

* Walter F. McCulloch, *Woods Words: A Comprehensive Dictionary of Loggers Terms*
(Portland, Ore.: Champoeg Press, Inc., 1958).

equipped with two husky strap irons on each side. They came down from the saddle close to the mule's sides, turned out about one and a half feet, then turned up, making a sort of basket. The wood was piled in the bottom and up the side, keeping the two sides balanced, then right on until the two stacks met on top. Packsaddles were padded with more than normal blankets and pads.

Mules were small, probably half burro. Nine or ten head were used in a string. They were broken to follow the leader without being tied together and as they came down the trail they gave the appearance of mobile cordwood piles. Arrived at the mill, the lead mule reversed ends, backed up to the fuel pile, scrunched down with his hind quarters until his hams almost touched the ground. The packer took his

Courtesy Calvin L. De Laittre, Minneapolis, Minn.
INTERIOR OF AN EARLY-DAY MINNESOTA BUNKHOUSE
NEAR ESQUAGAMAH LAKE
Loggers occupy the deacon seat, a plank along the lower bunk running the length of the building. Bunks were usually "muzzle loaders"; the men's heads were near the wall, their feet braced against the footboard, as shown on the right. Other bunkhouses had different arrangements.

place by the mule's shoulder and with a hand on the load on each side started a one-two rhythmic lift and on the third count gave a final boost and the whole load toppled endo back onto the woodpile. I would swear the mule, too, managed some upward thrust in front on that third count. Relieved, he straightened up, moved out of the way, and the next mule lost no time in getting into position to be freed of his burden. It was all done without a word from the packer.

Early-Day Western Montana

At a little later period logging in northwestern Montana had some variations from the Minnesota practices but horses were still in use. The country was far rougher than Minnesota so hauls might be down some steep hills.

Every camp, sooner or later, had the misfortune to have a load get away on the downhill run. The huge bunkload of logs, many tons of weight, would go tearing downhill with the horses dashing in front, trying to keep from being run over. If a horse ever fell, it was too bad. These horses had to be strong but quick and agile. The loggers usually contracted with farmers for their winter use, but some raised their own. The farmers couldn't normally use their horses in the winter and the logger couldn't use them in the summer, so many early-day logging contractors were farmers.

Jack says he remembers one outfit in Montana that always had a hunter whose job it was to supply the camp with venison. There were no deer limits or seasons then. In any case it was important to supply good food in incredible amounts. A day was ten hours then. Often the thermometer mercury got scared; it retreated into the bulb and forgot to come back for days at a time. Big men who work all day long at below-zero temperatures don't pick around with their food. Brute strength was a recognized and prized virtue. The admiring comment, "He's a good man," had nothing to do with morals.

In *Big Sam*,* a recent book about logging and loggers, is

---

* Samuel Churchill, *Big Sam* (New York: Doubleday & Company, Inc., 1965).

HORSE SKIDDING

The identity of the location has been lost. It is probably in northern Idaho.

a revealing incident. Fighting and drinking went along with the business. In a permanent camp lived Big Sam's wife, a dainty lady from Boston. She found two loggers fighting furiously. She stopped them and said severely, "Now, what are you fighting about?" They looked at her, mouths open in astonishment. One blurted, "My God, Mrs. Churchill, do we have to have a reason?"

Snow or rain didn't stop workers. The heavy work pants, long woolen underwear, and heavy German sox needed drying during the night, though some didn't remove the underwear until spring. To step into a bunkhouse at night with all this drying was an experience in pungency of odors. Those who didn't change underwear until spring gave off a distinct brassy fragrance.

"The Old-Time Lumberjacks" was written by Alice Hutchens who homesteaded on the old Blackfeet Forest.

"THE OLD-TIME LUMBERJACKS"

No story of the woods is complete without something about the old-time lumberjacks. When the days of logging with horses and of river drives were over, a lot of these old boys were getting along in years. Most of them never saved any money. They were too old to take up with new ways, so they went to the Salvation Army hotels to live, or to the County Home, or if they were lucky, they became caretakers at summer cottages and camps, or just "shacked up" at deserted cabins out in the woods. Most of these men were middle-aged when I was young. I doubt that one of them is alive now. Some of them were celebrities among forest folk.

There was Hughie, a little, wizened man who drove a four-horse sled team of huge matched blacks. He was the best four-horse teamster anyone around here ever saw, and he spent hours before and after work just grooming those horses. All his Sundays were spent with them. He had the usual failing of the old-time lumberjack—about so often he had to get drunk. At heart, Hughie was a gentleman—lots of those old-timers were. So when Hughie had been to town, Mr. Gardner, the woods boss, used to bring him home to the farm and put him to bed in the bunkhouse until he was presentable. On one of these occasions, he paid me a nice compliment: "Miss Jackman is a real lady. Just because a man's been foolish and spent all his money on drink, she will still sit at table with him and treat him decent like she does everybody." Of course by the time he "sat at table" he was all spruced up and ready to go back to camp.

There was Dan Styles who drove the crosshaul team, pulling logs from the "deck" to place them on the sled. He drove a team of two magnificent, intelligent dapple grays. He drove them without lines, just by talking to them, on a job where a slackening in the pull or a stop a few inches too soon or too far might have caused the death of the man who was top loading. Dan was a farmer in summer. I

Courtesy U.S. Forest Service
AN EIGHT-WHEEL WAGON IN MONTANA
The loading is being done with horses using a crosshaul, and the team can be seen beyond the wagon.

think he went to the logging camp in winter mostly because he loved handling horses.

All six of these fine horses (and several others) were poisoned the year the I.W.W.'s caused such havoc in the woods. Even yet, when I think of it, I feel like crying. Such a hideous thing—to take it out on those fine animals.

There was Jim Sullivan, who drove the first ox teams into the valley in the winter of 1909-10. They passed our house on 5th Ave. E. in Kalispell on their way to Swan Lake where a swamp was being logged and horses couldn't go. He was caretaker for the Tobies, up west of our home place, for years, until he was in his eighties. We saw him often and knew him well. He was Irish as ever could be and good-hearted to a fault, and given to getting himself into the funniest situations.

He came down one day looking very worried.

I: Is something wrong, Mr. Sullivan?

He: Well . . . yes, Mrs. Hutch, there is. I've lost me uppers. I put thim jist under me pillow as I always do, and this morning they're not anywhere at all.

I: Maybe you put them inside the pillowcase by mistake.

He: I shook the pillow, Mrs. Hutch, but they weren't there. They weren't anywhere. The Bansheen have got thim this time, for sure.

Next day he was down again and he was wearing the uppers.

I: And where did you find them?

He: Inside the pillowcase, just as you said. *But they were-n't there*, Mrs. Hutch. The Bansheen brought them back again, that's all.

It would have mortally insulted him if any of us had ever gone past his cabin without stopping in for coffee. It was a real ordeal. The pot was always on the back of the stove and

was often so full of grounds that there was scarcely room
for a cup of liquid.

He had a dog and a cat. He had taught both of them ele-
gant manners and all sorts of tricks. One time his roof
caught fire and the cat told him about it by alternately
clawing at his leg and looking fixedly at the stovepipe. His
theory about animals was this: You have to give animals
credit, you do.

One summer an old pal of his came to live with him. They
fell out over the silliest thing! One morning when they got
up to cook breakfast, there wasn't an egg in the house. They
waited until ten o'clock and the hens still hadn't laid, so they
had to make pancakes without an egg. Cooney got up from
the table, remarked that "a man couldn't even play solitaire
on a breakfast like that," packed up his things and left.

Of course Mr. Sullivan came right down to our house.

Courtesy Blister Rust Control, U.S.D.A., Spokane, Wash.
LOADING WHITE PINE LOGS ONTO CARS
A self-propelled belly slider loader is being used. Operation of the Ohio Match Company
in northern Idaho.

S: Well, Mrs. Hutch, Cooney's left me.

I: Oh, he'll be back.

S: Oh, no he won't. He took all his belongings this time.

But Cooney came back, bringing another old friend along. When the three of them got their old-age pension checks, they played a game of poker to see who should go to town and get the month's groceries and, of course, a bottle apiece. Cooney was low man, so he departed with his packsack and the three checks.

He didn't come back. The others accused him of all manner of ill-doings, and finally decided that, much as they had always trusted him and treated him so well and all, he must have taken all the money and gone off somewhere.

About two months later someone found him on a shortcut in the woods. There was one empty bottle, one nearly empty, and one full one, all the groceries, and all the change.

His friends mourned for him properly. He had proven himself honest after all.

Northern Idaho

Before and after the 1930's the Ohio Match Company had a sizable operation on the Coeur d'Alene National Forest in northern Idaho. It produced matchsticks, and used western white pine exclusively. They still used many early-day methods but practiced some more modern things. Falling and cutting into logs was done with crosscut saws. Skidding was with horses. Log chutes were in use to reach back farther from the railroad than horse skidding would permit. Chutes were made by placing two logs side by side with the face of each hewed to make a rough trough. They rested on short cross pieces at the ends, and other sections were added to make a continuous trough. An old chute builder could put them together so that any sized log would ride down the chute, around curves, like boxcars on the main line. A small crawler tractor would hitch onto the tail end

Courtesy Blister Rust Control, U.S.D.A., Spokane, Wash.

WHITE PINE OPERATION OF THE WINTON LUMBER COMPANY

The rollaway at the head of the lumber flume is at the extreme left. These were the short-log days in North Idaho.

log that would push twenty or twenty-five logs ahead of it.
The bearing faces of the chute logs would be daubed with
heavy oil or grease so the logs would slide easily. The grease
monkey used a three-gallon bucket and a swab on a stick,
and was adept at applying the oil as he walked.

The Ohio Match Company never adopted the flume as a
means of moving logs from woods to landing as some lumber companies did. The flume is more expensive to build
but cheaper to operate than the chute. The flume consists
of a V-shaped wooden trough resting on frame supports.
The two sides, of lumber, are about four feet high. Creek
water is diverted into the flume and logs are sluiced down to
the landing. Speed increases with the steepness.

At the logging railroad, logs were loaded onto flatcars with
a hoist. It rode on the car and propelled itself from one car
to another as each was loaded. Logs were small and were
picked up in bundles with a pair of cable chokers instead
of singly with tongs. The sapwood, less brittle than old
heartwood, was preferred for match handles. Therefore the
utilization was very complete and logs with top diameters
down to 4 inches were welcome.

Geared locomotives hauled the loaded cars over one summit and down to the Spokane International Railroad. Even
with these mountain-climbing locomotives, five loaded cars
were the maximum number taken over the steepest section.

The S. I. picked up the cars on their regular runs and
hauled them to Coeur d'Alene. Here they were pushed out
onto a trestle and the logs dumped into the lake. It was
quite an attraction to kids and grown-ups alike to see the
logs roll and see and hear the splash. They were kept in
booms until needed at the sawmill when they were floated
down the Spokane River a few miles to the mill at Huetter.

Here the logs were sawed into match plank and stacked
in the open air to dry and season. As needed, the planks were
again loaded on cars and hauled to Spokane to the cut-up
plant. Here they were sawed into sections exactly 2¼ inches
long. Women and boys armed with light hatchets then split

out all the knots, cross or crooked grain, pitch, and other defects. The straight, grain-perfect pieces were loaded into boxcars and shipped to Wadsworth, Ohio. Here the individual matchsticks were machine punched from the blocks and the sticks were dipped to make the head. The finished matches were placed in special boxes and the boxes into cartons. After shipping back to Idaho, we could at that time buy a box of Ohio matches for five cents.

Thirty years have changed things. Crosscut saws are replaced by power saws that are used for falling, bucking, and limbing, thus displacing axes too, in part. Big tractors have replaced horses for skidding. Any crawler tractor is called a "cat," although strictly speaking a caterpillar is made by

Courtesy U.S. Forest Service
TUGBOAT TOWING BRAIL OF LOGS ON THE ST. JOE RIVER
Near St. Maries, Idaho. The logs are being taken to sawmills at Coeur d'Alene, a distance of about fifty-five miles. Timber from many national forest sales are moved to mills by this method. The picture, taken in 1963, shows the tow running far back and almost out of sight.

only one firm. Whole trees are skidded out of the woods instead of "short logs" of sixteen feet. Topping and limbing is often not done until the tree reaches the landing.

Powerful loaders pick up the long logs and lay them easily on the trucks. Steam shovels were developed for dirt moving, excavating, and other construction work. Then gas shovels were developed. They were later modified for handling logs but the name stuck. The log loader is usually called a shovel.

Cats were equipped with a dozer blade and bulldozers revolutionized road building. Roads are now found where no one dreamed that logs would ever be trucked out of the woods.

Roads brought more and better trucks and the shovel loaders, able to handle long logs, made necessary bigger and longer trucks. Long reaches were required. The piggyback was invented so the rear assembly of wheels could be pulled up on the truck proper and ride instead of roll back to the woods for another load. Eighteen big tires are commonly used on each truck. A fifty-mile haul is not unusual. Once the truck is loaded, out of the mountains and on a good road, a few more miles makes little difference. So we see loads of logs going north, meet other big loads going south.

Donkey engines and high lead cables were used predominantly in the big timber of the Pacific Coast, but this system with its topping of the spar trees never spread to our inland country. Even that method, so popular on the Coast, has given way largely to cat skidding, shovel loading, and truck hauling. The latter is more mobile and requires no railroad operation.

Free Use—The Farmer's Woodlot

When the national forests were established, dead timber was made available without charge for farm or other noncommercial use. The need was urgent for posts, poles, house logs, fuel, and dozens of other things, such as doubletrees and chopping blocks. The Minidoka Forest in southern

Idaho had five divisions and each division was a separate
mountain range. The surrounding land was settling up. First
came livestock ranches, then dry-farm homesteads. Dams on
the Snake River were providing irrigation water for many
settlers near the river. Initially free use permits were issued
and we had 2,500 families getting these minor forest products from that one national forest. They came with team and
wagon, bedrolls and camp outfit. It was an outing, too, and
sometimes the family rode along on the running gears. They
got away from hot dry weather at home. Often it required
two days to come and two days to return. Besides cutting and
loading their material they sometimes caught a few fish or
picked wild berries, or just enjoyed the change from the
gray sage to the evergreen country with its shade and cool
nights. They needed corral poles, posts, barn logs, potato

Photo by Bob Bailey, Enterprise, Ore.
LONG LOG SKIDDING IN THE PONDEROSA PINES, OREGON
A recent picture

cellars, fence poles, chicken roosts, bean poles, hogpens and other things of wood. Here they were able to get what they needed, spending only their own labor. Cash was scarce. After a few years permits were no longer required, which saved time for forest officers and users. Removal of dead timber reduced the fire hazard. Lodgepole pine was one of the main species and in thick stands the slower growing trees were overtopped by the dominant ones and died. Some were as small as tepee poles and others were all sizes up to house logs. They were straight and had little taper, and folks were glad to get them. They lasted for a couple of generations if kept from touching the ground.

Later, following the Big Cow fire of 1939 on the Malheur and Whitman Forests in Oregon, a somewhat similar need developed. A portion of the fire covered a lodgepole pine area. It killed everything but was not hot enough to consume the timber. New irrigation projects were under way at Vale and other points near the Snake River in Oregon and Idaho. Potatoes were a favorite crop but the new settlers couldn't build concrete cellars. They flocked into the burn at the head of the South Fork of Burnt River and salvaged several million feet of fire-killed timber. Cattlemen from as far away as Nevada came to the burn for corral poles. The Forest Service was hard put to open up passable roads to keep ahead of the demand. Quite a number of the new settlers around Vale and Nyssa in eastern Oregon were Japanese, and Japan at that time was our enemy. Loggers working for a lumber company cutting fire-killed yellow pine sawlogs not far away objected to the Japanese coming onto the forest for government timber. They even threatened to do bodily harm if the practice continued, and they could not be sidetracked. The Japanese were understanding and the difficulty blew over in a few months. These Japanese are some of our best citizens, helping on community work of all kinds, Americans now.

Christmas trees for personal use were free and those who liked to load up the kids and go out to cut their own were

VIEW, FROM McDONALD LOOKOUT, OF A FINE STAND OF WHITE PINE TIMBER ON STEAMBOAT CREEK, COEUR D'-ALENE NATIONAL FOREST

Timber can grow like this only if there is soil to support it

welcome. They were asked to take the trees where the cutting would serve as a thinning and thus improve the condition of the trees left. Of course, the choice, bushy, regular-shaped trees developed in the open and who would want a spindling or lopsided tree growing in a clump? In late years increased demand and more public concern for roadside appearances have required restrictions and more control.

Wood for fuel has been removed in huge quantities. During depression days, or unemployment, many people fell back on wood and hauled it themselves. Wood stoves for both cooking and heating have been thrown out now along with churns, buggies, and milking stools.

S-22, Sales at Cost or Farmer's Sales

From early days, farmers and ranchers could get timber from a national forest, for use on farm or ranch, by paying the cost of handling the sale. We charged a flat rate of seventy-five cents per M board feet. We selected trees and marked them for cutting. The logs were scaled (measured, and volume determined) or often the volume of each tree was estimated and payment based on that. The farmer did his own logging and hauled the logs to a sawmill where he had them custom sawed. The intent was that the farmer would have all the products from the trees that he bought, including slabs, edgings, and clear boards. In practice the farmer perhaps could use lower grades just as well as the uppers and the sawmill operator was prone to trade for the clear boards and give a bonus. It was taboo for the farmer to pay his sawing bill with some of his lumber, but this was often a temptation if he had more boards than money. The purpose of this authorization was to give the farmer stumpage at a nominal cost and it served a useful purpose. Many a ranch house, barn, granary or outbuilding was built from seventy-five-cent stumpage. As the pioneering period passed, sales at cost fell off and were discontinued.

Commercial Sales

On most forests in the early days, small sawmills in the woods, close to the timber, were common. They used a circular saw and the carriage tracks were not always in alignment or fully stationary, so the resulting lumber was irregular in thickness. The owner of a little mill had a hard time buying the mill and had no planer. Rough boards and dimension were the product. Slabs and dead wood were burned to produce steam for power. A stream of water sometimes carried the sawdust from the mill back into the creek. Usually an open fire or makeshift burner disposed of surplus slabs or sawdust. Their insurance rate was exorbitant so most of the owners didn't carry any. Fifteen or twenty of these mills would be operating on a single forest. One district ranger, traveling horseback to scale the logs at one of these mills, was in the habit of leaving his scale rule at the mill. It was consumed in a mill fire. Forest officers were held accountable for all nonexpendable government equipment and used Form 858 (disposal of government property)

A SMALL CIRCULAR MILL RUN BY A STEAM BOILER
FIRED BY SLABWOOD
Hundreds of such mills served their purpose and went by the board

to get relieved in case of loss. On the 858 this ranger stated, "I left the scale stick at the sawmill. The mill burned down and the scale stick burned up."

Such mills have been replaced by a few large modern mills. Fast band saws make quick work of a big log. Electricity provides power and everything moves on rollers. A real push-button job. Modern planers turn out surfaced material to exact specifications. A cut up plant in connection may produce molding, siding, window and door casings, flooring, trim and furniture parts, or box material. Dry kilns change the "pond dry" lumber from green chain into fully seasoned lumber in a matter of hours. Rossers remove the bark from a revolving log. The barkless slabs are chipped for pulp. Waste burners are still in use but they consume a much smaller percentage of the log.

A mill utilization study in 1925 showed that the average western white pine tree logged in North Idaho was 23 inches in diameter breast high. Only 64½ percent of the average log went into lumber, 10 percent went into slabs, 20 percent went into sawdust, 3 percent into edgings and 2½ percent into trims. Thinner saws have reduced sawdust. Former waste material is largely used. Plywood, pulp products, wallboard, pressed wood and particle wood have replaced boards for innumerable uses. Further, plywood makes no sawdust or slabs.

Plywood has taken over in a big way. At first only large clear Douglas fir logs, known as peelers, were considered suitable for plywood. Now lower quality logs of several species produce much of this material. It is stronger than boards and takes less labor when used for sheathing, roofing, subfloors and such purposes. With the new waterproof glues, it is even used for boats.

From the beginning Forest Service sales in excess of $500 had to be advertised for thirty days. Of course $500 went quite far with early-day stumpage prices and a small circular sawmill. Sealed bids were opened on the date and hour designated. This system gave all bidders an even break, but there was still a large degree of chance in it for the operators. Sometimes the most logical purchaser lost to a newcomer by only a few cents per thousand feet that he would gladly have paid. Sometimes a local operator, knowing that others had been looking over the timber, bid perhaps a dollar a thousand above the advertised price only to find that he was the sole bidder.

In the late forties a change to oral bidding was made. Date and hour for sale are published as before. Sales under the new plan do not resemble a farm auction where the hammer threatens to fall any minute and bidders must decide quickly. The air can be pretty tense where two or more bidders really want the timber and the total value sometimes runs to six digits. A blackboard is used to record bids. An adding machine may be needed when several species are involved with differing appraised minimum rates, because the total valuation counts. Scores of raises sometimes are made. Time is allowed for conferences, phone calls, or computations. Finally, after it is apparent that no one wishes to make another raise, the last bidder is declared the buyer. Of course anyone, to qualify as a bidder, must make a stumpage deposit to show ability and good faith. Deposits are returned to the unsuccessful bidders.

The Forest Service has been criticized by the industry for charging exorbitant stumpage prices. An appraisal is always made and a minimum price established. Only rarely have there been no bidders. The really high rates have resulted from competition between bidders where one operator figures he must get the timber to continue in business or another is determined to break into a territory new to him. Modern transportation extends the operation range of all mills.

Sale contracts were entered into even in the small unadvertised sales. Payment for stumpage was required before trees could be cut. Installment payments could be made. Occasionally it became necessary to shut down the falling crews until some little operator made another stumpage pay-

ment. Maybe he couldn't make a payment until he sold another load of lumber. Forest officers were lenient on many
scores, but this was one place where money talked.

Payment was based upon the measurement of the timber
cut. Scribner Decimal C was the standard scale and the
rule had burned on it the volumes for all diameters and usual
lengths of logs. A scaler came to know all the scales for all
sizes of logs. The only one I remember this minute is that a
16-foot log, 16 inches in diameter (inside the bark at the
small end) scales 160 board feet. A scaler needs to be a combination tap dancer, spider, and water ousal. On some jobs
he measures and marks the logs in the woods behind the
fallers. Sometimes logs must be scaled on the trucks. Sometimes it is best to scale in the mill just before logs go on
the carriage. They may be scaled in the water. In this case

Courtesy U.S. Forest Service
PONDEROSA PINE THICKET, WHITMAN NATIONAL FOREST, 1929
"Dog hair" to hunters and loggers. Nature will thin this out after a long waiting period

experience counts for the logs are mostly below the water line.

Circumstances govern. Are other ownership logs coming to the mill? Are government logs from two sources (two prices) being mixed? Where best can defects, lengths, etc., be determined? Is the operator hot logging, decking, cutting from two or more sides or faces? In farmers' sales or real small commercial sales, selling by tree estimate was practiced as the only economical way. Diameter breast high, "D.B.H.," was determined, by caliper, diameter tape, or Biltmore stick, and the number of sixteen-foot logs estimated. This gave a reliable figure except for invisible defects that should be deducted from the gross volume. Where logs of private ownership are mixed in river or lake booms, each owner usually

Courtesy U.S. Forest Service   *Photo by Fred Furst*
THE SAME PONDEROSA PINE THICKET AFTER IT HAD BEEN
THINNED BY FORESTERS
The thickets, without thinning, produce stagnated stands for forty years or more.
Thinning releases crop trees that grow fast into salable timber. Near Susanville, Oregon,
in 1929.

brands his logs on the sawed surface at the end much as cattle are branded. Each owner has a different brand.

Timber Management

The marking axe has been described as the instrument of the forester. The trees removed govern the composition of the remaining stand and kind and amount of crop to be harvested in the future. Research men have provided general instructions for marking timber for cutting but the forester swinging the marking axe determines the outcome right there and then. Every tree is different and a thousand decisions must be made in the day's work. (Like the potato sorter who quit—too many decisions.) Nature isn't systematic, conditions vary, dozens of factors influence decisions. Anyone can swing the axe, but it takes training, experience, judgment, understanding of influences, and a real conception of ultimate goals to be a topnotch timber marker.

Ponderosa pine characteristically grows in uneven aged stands. On a sample acre you may find a big flat-topped, red-barked, overmature veteran. Two or three younger but fully merchantable trees make the fallers' mouths water. A larger number of bull pines would be scattered about. (If you really want to start an argument with an old-time lumberjack just say that a "bull pine" is a yellow pine or will turn into a yellow pine in another fifty to seventy-five years.) You would probably find a sizable area of yet younger trees in large- and small-sized poles. In an opening where a real old grandpa had been blown over or killed by lightning or bugs, there would be seedlings maybe a foot to ten feet high. This makes an ideal setup for selective cutting. Trees left respond remarkably to increased sunlight and moisture. Reproduction is already established. The openings created are readily reseeded by nearby trees. A second cutting can be made in far less time than is required to grow merchantable-sized trees from seed. The cut might vary from a light culling to one of 60 percent or 70 percent by volume. Alex

Jaenicke (timber management specialist) used to tell us to leave trees for "safe storage." They would be in the second category of our sample acre, fully merchantable now, but thrifty, sharp topped, and rapidly increasing in volume.

Inland and Rocky Mountain Douglas fir and western larch are not so well adapted to selective cutting as is yellow pine. But they are often intermingled with pine, or in small pure stands on north slopes. Since pine is the more valuable species it is favored at the expense of the mixed species when marking for cutting. And there is enough spread in thrift and age classes that a somewhat modified selective cutting in the fir and larch works well.

White pine, within any small area, occurs in even-aged stands. One area of young poles six or eight inches in diameter breast high may butt up against a stand of old overmature pine. A mile away it might end abruptly at a stand of healthy sixteen-to-twenty-inch pines. This is the result of local fires occurring years ago. White pine sometimes grows in mixed stands with white fir and western hemlock. Each species may represent about one third of the stand. This creates a silvicultural puzzle. White pine with its soft, easily worked grain and high durability is greatly prized. It is in demand for pattern stock, window sash, trim of all kinds, drop siding, and match stock. Hemlock and white fir were known to make good pulp, but there was no demand whatever for it in our territory in the thirties, and these trees weren't worth cutting for lumber. Both white fir and hemlock are prone to defects at a relatively early age. The mature trees are liberally adorned with big fungus conks. At best they have a rotten heart. At worst they have only a thin shell of sound wood. Choice white pine of good size are found in these old-growth stands with fir and hemlock. But if the white pine is removed the chances are about ninety-nine to one that young fir or hemlock seedlings will fill in the openings.

Confronted with these conditions an unorthodox sort of cutting practice was followed: a little clear cutting with numerous seed trees left, some selective cutting, some thinning, always having in mind the goal of favoring white pine at the expense of its unwanted sisters.

About 1936 a revolutionary step was taken on the Coeur d'Alene. We laid out tracts of these decadent mixed stands and sold, for cutting, *all* the white pine. After the pine was removed we put in fallers whose sole job was to lay down every remaining tree on the area. Where normally the white pine slash (tops and limbs) was piled and burned, here it was left where it fell. A cleared fire lane eight or ten feet wide was made to separate the slashed area from thrifty green timber adjoining. In the fall the cutover areas were broadcast burned and the following spring white pine seedlings were hand planted.

Those controlled burn jobs were the most intriguing imaginable. Enough risk and gamble were involved to suit anyone who liked to live dangerously. Yet you could choose your time and weapons. We burned only at night. The material had become tinder dry. The ideal time was four or five days following a quarter-inch rain. The duff under the green trees outside the fire lane would still be damp, but the slash would be dry. We would assemble a picked crew of experienced fire men. Tools were assigned, including some pack sprayers for putting out spot fires and plenty of Houck or other propane torches. Each man had his specific job. Most of the areas had rather steep slopes. The firing started at the top. When everything was ready, everyone in his place, a couple of top hands began gingerly to use the torches to start the fires right at the point. If there was any breeze the fires would be started on the lee side. After they got the feel of it more fire would be spread, a few yards inside the lane and a short distance down each side. Time was allowed for these starts to work downhill a bit and to consume the main flash fuels, widening the built fire lane. Later, at a signal, more torchmen started fires across the area and about a third of the way down the slope. These fires, when they got under way, really took off and in a short time ran into the now dying fire above. The flank torches had slowly worked down each side, thus adding a burned strip to the cleared fire lane. When the heat was subsiding and things getting monotonous, the materials all along the bottom were touched off. Then the real fireworks took place. Everything was under control. By morning stumps and some logs would still be burning but the danger was over.

This practice was followed only where fir and hemlock were old and badly defective. Young thrifty stands were not sacrificed, but no one then dreamed that thirty years later fir, at least, would be in good demand.

Lodgepole pine, while much less valuable, covers extensive areas. Even forty years ago (1920-30) it was in localized demand for power poles, telephone poles and minor uses.

READY FOR A CONTROLLED BURN
On clean-cut areas all white pine timber was cut and removed. See page 132. Then all hemlock and white fir was felled on top of the white pine limbs and tops. Only over-mature and badly defective hemlock and fir areas were so treated. Note the log chute down the draw and the cleared fire lane by the green timber.

In Utah it is known as "bird eye pine" from the markings just inside the bark that resemble bird's-eye maple. Lodgepole is definitely a "fire" species, and clean cutting with natural reseeding is indicated unless one wants to bet that he can beat nature by trying to convert to some other kind of tree. Lodgepole cones tend to stay closed until a fire occurs, when they open and scatter their seeds. If any are in a stand, a burn kills the other species and the burned area comes into lodgepole, maybe making a dense, tangled thicket much cursed by hunters who try to fight through them, not knowing that the thicket may be a couple of miles across.

One group of local citizens that really appreciated lodgepoles was Indians. They used the long, slim poles for teepees, travois, burial stands, frames for treating hides, poles for meat racks. Use was limited only by the imagination. Most of these uses have been wiped out by civilization.

Sustained Yield

"Sustained yield" is a phrase tossed around freely these days, both in print and on street corners. I suspect many who use the term would have difficulty giving a clear definition of it. It has no relation whatever to present sawmill capacity. The mills in a community may be able to handle one hundred million feet of timber per year. Yet it is conceivable that only half that much timber can be cut annually on a perpetual basis.

Sustained yield is tied in with timber growth, yet it isn't wholly correct to say that the sustained annual yield equals the annual growth. If in any block a preponderance of the timber is old and decadent the current growth is low. But unless the old growth is too far gone, it can be used to provide a larger cut than the actual growth would warrant. But if a preponderance of the unit consists of only seedlings and saplings, timber ripe for cutting cannot equal the annual growth during the fore part of the cutting cycle. The loss of timber due to fires, insects, disease, and windstorms has a bearing. The forester can either make a long guess as to the probable loss that will occur and establish his sustained yield figure accordingly, or he can disregard this factor. In this case he can only say he is sorry if calamity proves him to have been too optimistic. Besides, he may be gone then.

Changes may require revisions of the sustained yield—for example, a new use or method. The figure for the Whitman National Forest now is thirty million feet more annually than it was fifteen years ago. Extensive areas of rugged terrain were formerly classed as nonoperative and timber there was not included in the calculations. Everyone thought God had provided those areas for the billy goats. Modern dozer manufacturers and present-day cat skinners didn't hear about the original plan. Fifteen years ago white fir logs cut on road construction on an operator's own land were pushed over the bank as not worth hauling to the mill. They are now in demand.

Forestry professors use this hypothetical example: Suppose you have a hundred acres of timber-growing land. One acre has one-year-old trees, the next acre two-year-old trees, and so on to one hundred years. Assume that a tree is ripe for cutting at one hundred years, which may or may not be true. Assume you have no fire losses and no blister rust, no tussock moth and no spruce budworm attacks. Suppose an acre of your ground will produce nine thousand feet of timber in one hundred years—your sustained annual yield is exactly nine thousand feet per year.

Stand Improvement

In the early timber sales, provision was made for a small deposit into the "Co-op Fund," for stand improvement. At first when stumpage was low at two to three dollars, only ten or fifteen cents per thousand feet went into this fund, but with later increases in stumpage a corresponding increase into the co-op was made. This was taken into account in the appraisal and the stumpage price would have been just that much higher if no funds had been collected for stand improvement. The idea was that when government timber is cut, it should help to some extent to leave the area in good condition for another crop. Each sale provides funds to help improve the remaining crop.

Cutting out part of the saplings from thickets, known as thinning, is one form of stand improvement. Nature's way is to start many more trees to the acre than can survive. Gradually leaders get ahead of their neighbors. They get more than their share of sunlight and soil moisture. The suppressed saplings eventually die. In the process the leaders or crop trees are retarded and a fifty-year period of virtual stagnation results. Man-made thinning eliminates this competition, produces faster growth and a uniform spacing of remaining trees. Yellow pine responds well to these release cuttings and shows remarkable increase in growth. The Whitman Forest reports 2,262 acres thinned in 1964. Thinning is expensive but if started now on a massive scale the nation would be repaid well.

Pruning is done on some sales areas. A light curved saw with handle twelve feet long is used. Trees four to seven inches in diameter are selected and all limbs removed to about eighteen feet. This means a sixteen-foot log will provide clear lumber instead of knotty boards. Clear lumber in the past came largely from big trees 250 to 300 years old that eventually had shed their lower branches permitting growth free of limbs (knots in boards). The inner parts of these big clear logs usually have knots in them, showing where limbs had been in the trees' youth. Pruning of small trees insures some clear future lumber even from moderate-sized trees.

Slash Disposal

In early days timber purchasers were expected to pile and burn the tops and limbs from trees cut. The operator usually was too busy trying to cut enough logs to keep the mill running. Slash disposal was sort of a necessary evil to him. And what lumberjack wouldn't quit if he were sent out into the woods to pick up limbs? That was one step lower than a swamper, who cuts off the limbs.

So a change was made whereby the operator paid into a co-op fund that could be used by the ranger to dispose of the slash as a Forest Service job. In the white pine we used a figure of seventy-five cents per thousand feet of timber. Piling was done during the summer in small compact piles that were burned after about two inches of rain had come down in the fall. A foreman and small crew on brush work all the time became efficient and were ready hands for first call on fire. Howard Drake, Timber Management staff man on the Coeur d'Alene, developed a gyppo plan of slash dis-

Courtesy E. R. Jackman, Corvallis, Ore.
SELECTIVE CUTTING IS PRACTICAL IN PONDEROSA PINE
This species responds remarkably with increased growth following selective cutting (the removal of ripe trees). The immature trees grow rapidly and the seedlings come in on the openings.

posal. He made a verbal contract with each of several reliable experienced foremen to pile a certain block for a figure
slightly below the seventy-five cents the purchaser of timber
was committed to pay. The gyppo hired his own men, fed
them, and supervised them. Periodically he would be furnished a statement that he had a certain amount earned and
the timber purchaser paid him direct. We were relieved of
payrolling, leave records, accident reports, and other details.
Finally the auditors held that we were spending undeposited
co-op funds. So we went to formal contracts let to the
lowest bidder.

Irresponsible fly-by-night bidders cut the rate and, after
getting the jobs, tried to cut corners, had labor difficulties,
and were so much trouble that we had to go back to day
work. The actual slash disposal work hasn't changed much.
Piling the limbs and tops while green and burning the piles
after fall rains started has usually been the practice. A variation has been used in some places. Strips along roads or trails
or other higher risk areas are cleaned up more completely—
that is, we dispose of not only slash from timber cutting
but windfalls, dead snags, and all really inflammable material. A few lanes through the cutting area are cleaned up
but the major part of the actual logging slash is left undis-
posed of. The object is to reduce the fire hazard following
logging as much as possible with a reasonable expenditure.

Road Construction

In the early years and until the 1940's a purchaser of timber
built the roads he needed to remove the timber and he built
them to last only until he could get his logs out. Some were
so steep they washed and gullied badly. Instead of culverts,
rotten logs or dirt fills were used. In many cases the roads
became impassable in a couple of years. Drainage is the most
important single feature of a road and this is doubly true in
mountain country. Without a great increase in first cost

a road can be so located and so built that it can be used indefinitely with reasonable maintenance.

Gradually higher road standards were used. The purchaser builds the roads but the estimated cost of road construction is used as an expense item in working out the appraisal of stumpage value. A higher road cost means a lower stumpage price. The purchaser is not penalized by building a more lasting type of road. The Forest Service makes the survey, or approves the location and prepares the specifications. These include grade, width, number and size of metal culverts, curvature, sloping of banks, and disposal of right-of-way debris. Instead of verbal agreements for temporary use roads across private tracts, permanent easements are now secured. The timber builds the road. The permanent road enables later harvesting of windthrown timber, bug-killed or fire-killed timber or a light second cut. Of course it facilitates fire protection and fire fighting and, incidentally, hunting, fishing, and camping.

Planting

Planting is an important part of timber management. In comparison with the total area of any forest, the acreage planted each year will be relatively small. For example, the Wallowa-Whitman, with a national forest acreage of 2,223,426 acres, planted 2,540 acres to trees and seeded 261 acres with tree seeds in 1964. That is a little over 1/10 of 1 percent. Planting is expensive and isn't done wherever natural reproduction can be induced. Conifer seedlings are sensitive and must be well cared for and carefully planted. Rodents are detrimental and livestock and big game animals trample and disturb the trees. Drought before the seedlings get well rooted can be disastrous. After a burn the black ashes hold the heat and the soil can get hot enough to interfere with growth.

Planting is indicated on most clean cut areas, especially with controlled burning of defective old growth hemlock and white fir. Planting of accidental burns enables a more prompt stocking of conifers and of the most suitable or valuable species.

Tree planting seems to have a romantic appeal and catches the imagination of many persons. Schoolchildren become interested in fire prevention and forest conservation through tree planting. Even the Congress of the United States is a bit partial to tree planting.

CHAPTER SEVEN

SURVIVAL: Keep Cool and Live

What To Do If Lost in the Forest

ONE JOB THAT COMES TO EVERY FORESTER SOONER OR LATER is that of helping to find lost persons: hunters, hikers, picnickers, and children. In a slightly different classification is helping hunters or others out of the mountains when snowed in. Another is searching for and helping remove fliers who have died or been injured in plane wrecks in the forests. These things aren't taught in the forest schools. They aren't provided for in Congressional appropriations. They aren't listed in project or job lists. But they occur every year and the forest ranger, until recently, was the first to be called. He knew the country, the local people, the trails, the telephone stations. He usually had saddle horses and pack animals, and sometimes helpers. Being public spirited and somewhat the host—he tackled the job. Any reasonable expense was paid by Uncle Sam out of some general fund not strictly earmarked. The use of funds and the forester's time took just that much away from planned activities.

The sheriff's office got in on the special jobs, too. More recently useful organizations have grown up, such as sheriff's posses and volunteer search and rescue groups.

Ranger Gerald Tucker,* on the Umatilla National Forest in northeast Oregon and southeast Washington, had a major part in the rescue of forty elk hunters. They were snow-

---

* "Buried for Thirteen Days in a Blizzard," *Outdoor Life,* January, 1964.

bound for thirteen days in the Blue Mountains in November, 1945. The rescuers risked their own lives and put up with the most strenuous and difficult situations imaginable. An expert and self-sacrificing dozer operator really saved the day. The U.S. Army at Walla Walla sent in two rotary snowplows. Ranchers and other hunters cooperated.

A few years after this a group of elk hunters went into the same area and got trapped in the same way. The number of men was smaller and no blizzard was encountered. The party requested the Forest Service to send in a bulldozer and trucks. They had been advised not to go down on the back side of the mountain at that time of year. Supervisor Ewing sent them enough long-handled shovels to equip all the hunters. This time it warmed up and they were able to get out under their own power.

Plane crashes in the woods are usually bad. The planes hit the trees before they reach the ground and often mow off treetops for some distance. I have had to help find and bring out the dead from two crashes. It is for others than foresters to try to reduce airplane accidents.

We can offer suggestions that could result in a reduction of the injuries and deaths due to getting lost in the forests.

Some people do such crazy things when lost. They seem to lose all reasoning ability and go into blind panic. Jackman tells of one incident he knew about in Montana. An employee of the Great Northern Railroad was out hunting. Weather was good. When he didn't return, search parties went looking for him. The railroad company ran an engine up and down the tracks all night for several nights blowing the whistle. He wasn't found until the following spring. He had thrown away his gun, his fur cap, and his mackinaw. He had fallen across a log close to the track and died there. Had he just sat down and built a fire, rescue would have been swift and certain in his case.

Preparation

We are talking here about survival when one becomes lost in the forest. But actually it's better not to get lost in the first place. As in the sage advice in the old French-Canadian poem,

> You can't get drown on Lac St. Pierre
> So long you stay on shore.

Assume you are going out to hunt elk or deer. First I would recommend that you carry a light knapsack. One of the army type is big enough, is not heavy, and keeps necessities together. It rides well up on your shoulders and carries better than if the stuff is in pockets, tied on the belt, or slung around the neck. Besides, you don't look like Santa Claus. This should be an individual item. A party of two or three may plan to stay together, but hunters don't walk hand in hand. So each one should be prepared as if he were going alone.

The most important single item is plenty of matches in a waterproof container. A wet match is no better than none at all. Rubbing them in the hair gives them an oily film that tends to prevent absorption of moisture. Use the old-fashioned wooden ones. A commercial match holder is good, or improvised containers will do. A shotgun shell with the end plugged can be used. A small bottle with a cork or a tin container with screw cap will carry more. It is wise to carry matches in two places just in case. A "Marble" holder in the pocket and a larger supply in the knapsack is good. *Never* leave camp or car without plenty of well-protected matches.

A lightweight hand axe would come second. It should be sharp and covered by a sheath. That protects the edge and also the knapsack and its other contents. I have known a few who traveled without the sheath, but what if they slipped on an icy hill and got cut when they fell? Such a hatchet isn't equal to a double-bitted axe for heavy chopping but is mighty handy for cutting firewood. It's much better than a jackknife for cutting kindling, and chopping pitch from stump or snag. Other uses are: blazing a line from dressed game animals to a trail; marking a spot on road or trail that you might want to find later; blazing a line from an emergency camp or injured partner; dressing out an animal. An elk must usually be quartered.

A compass is very valuable. A small, wooden box compass is good. I carried the same one for forty-eight years until I loaned it to a supposed friend. One with the full circle 0 to 360 degrees is preferable to the N-NE, S-SW variety. But unless you are familiar with the compass and know how to use it—it isn't worth carrying. Then, too, trust your compass. A person who is turned around in his directions will swear his compass is wrong. Most compasses have one end of the needle black and one end silver. It is easy if you are excited to forget which end points north, and then it is most confusing. To avoid this, cut on the wooden box or scratch on a metal compass, "Black=North." You will refer to that more than once.

A map covering the area is most valuable. Don't carry a big 4'x4' map folded up. Get as accurate and detailed a map as you can and cut out a piece six inches square covering the areas you plan to be in. Be sure you understand how to read a map. Contour lines are helpful but can be confusing. Be sure of the symbols for water, trails, roads, cabins, telephone lines.

A good-sized clean square of white cloth has many uses. Some are: bandages; tourniquets; sling; protection as a neckerchief; signaling. A regular handkerchief is too small and has its own purpose.

Flashlight. A two-cell flashlight in good condition may be invaluable. You don't even have to be lost. In your concentration on the hunt you may get too far from camp. Or your watch may stop and darkness surprise you. Like the maid who was alibiing for the overdone eggs. She said the kitchen clock was slow. If you don't put a flashlight in the knapsack you may wish you had.

If you pack a lunch, the knapsack is the most convenient place to carry it. If it develops that you do lie out a couple of nights, a bit of emergency food helps to while away the time.

An extra pair of good wool socks and a spare pair of gloves don't weigh much and can be a real lifesaver. In case of wet feet a change to dry socks is a great comfort and may prevent some blisters. In cold weather and deep snow dry socks have saved feet from freezing. If you sit it out around a fire the dry ones come in handy while the wet ones are drying. Dry gloves can prevent frozen fingers.

First-aid kit. This is debatable. A kit is fine in car or camp. Ordinarily I wouldn't recommend taking it along in the knapsack. In rattlesnake country and in snake season, a venom kit, yes. If you are a known bleeder, a coagulent, yes. If any heart weakness, your prescribed remedy, yes, or better yet, stay home. A first-aid kit in camp doesn't do much good if you are five miles away and injured. But you can't take everything on your back you may need if you break a leg or shoot yourself.

Extra clothing. This depends partly upon weather and time of year. Possibilities are a sweater or a sweat shirt, light wool or leather jacket, or waterproof jacket. Often a person starts out when the weather is nice, then after he gets warmed up, the weather changes. It snows or rains, or the wind gets around to the northwest, or night sets in. He doesn't foresee the need when he starts, but would give a week's wages for some more clothes before he's out of the woods. So pick out one item you think most likely to be useful and put it in the knapsack.

Half of an ordinary tallow candle helps fire building if it is windy or the fuel is damp.

A dozen uses might be found for a pair of leather shoelaces.

This makes twelve items:

Knapsack	White cloth
Matches	Leather shoelaces
Hand axe	Flashlight
Candle	Lunch & emergency food
Compass	Socks and gloves
Map	Jacket

The total weight is under seven pounds. With them you're pretty well equipped to find your way out of the woods, or to stay there several days if you must.

Keep Your Head. Don't Panic

Some persons will panic as soon as it dawns on them that they are lost. There is a queer psychological reaction. Many start to run. They decide they are going the wrong way and turn and charge back the way they just came. They throw away guns and even part of their clothes. The panic obliterates all civilized traits. It changes a man into a crazed animal. His mind doesn't function. He will run right across a plain trail or even a road and never see it. He will splash across a creek, making no effort to keep dry, or to take a drink even if he is thirsty. He stumbles and falls over rocks and logs that he would ordinarily avoid. He tears his clothes to rags on limbs and brush; he cuts and scratches and bruises his own body without realizing it.

Jackman tells of a case where a party of his friends were camped at the edge of a meadow. A lost man who had thrown away his gun and outer clothes rushed by their camp only a few yards away. They called to him but he couldn't hear. Two of the men caught him, threw him to the ground and restrained him until some measure of reason returned.

Two hoodlums from Chicago killed a policeman in La-Grande, Oregon, when he was about to tell them that one of their automobile lights was out. They had stopped at a filling station for gas but hurriedly drove on toward Pendleton without the gas. Their car died near Meacham and they took to the forest. They were well armed but came out in three or four days completely terrified. They were worn out, gaunt from hunger and loss of sleep. Unfamiliar noises had made sleep impossible. They were overjoyed to find a human being and asked to be taken to the nearest police.

Panic is fear. Fear of wild animals, fear of suffering or starving, fear of embarrassment or ridicule, fear of the unknown. Some persons may not be able to avoid panic. If one knows that to be the case he should not go into the forest. Most everyone can avoid panic. First he must determine to keep his reason. The unknown always seems more sinister than the known. Therefore he should try to become familiar with the conditions, the problems, the ways to do

Photo by Bob Bailey, Enterprise, Ore.
CONTENTS OF AN EMERGENCY KIT
Take the kit with you, rain or shine. A checklist is on the opposite page. The packet
is a man-sized waterproof windbreaker.

things in the forest. He should go into that forest with a companion and spend several days looking it over. He must develop confidence in himself. The most practical thing if confusion starts is to SIT DOWN—relax, calm down, think —easy does it. Just hold the horses!

Keeping One's Bearings

This is just the opposite of getting lost. Most everyone who has spent much time in the forest has been lost. Jackman says he has been dozens of times. I've been lost, too. I was lost once and didn't know it. I was elk hunting in the Shaw Mountain area, Union County, Oregon, and traveling alone along the main ridge. There was new snow, good tracking depth. The sky was heavily overcast. From a saddle I contoured around the north side of a height of land. I was concentrating upon spotting game and had no concern about my directions. After about a mile I encountered a man's tracks headed in my direction but he had taken to the sidehill. I was aggravated that my intended hunting area had been worked ahead of me. But I continued along the opposite direction of the tracks, remarking to myself what big feet the guy had and what a long reach. Finally I came to a good-sized house rock and beside it the hunter had stopped to make a few extra tracks. The rock looked familiar and suddenly it came to me that I had made the tracks myself. I had followed around the sidehill on the north side, crossed over the divide in a low spot or saddle, continued on around about on grade but on the south hillside and headed west until I met my own tracks going east.

I can excuse myself for not keeping direction by the sun because there wasn't any. But it was absolutely inexcusable to overlook the timber types. On the north slope were Douglas fir and tamarack, young growth and poles, while the south side had open yellow pine. Any woodsman should have known he couldn't be going east through a south slope type with the high ground on his right.

SOUTH SLOPES ARE MORE OPEN, LESS BRUSHY, HAVE FEWER
WINDFALLS, ARE BETTER LIGHTED AND EASIER TO TRAVEL

An experienced woodsman keeps his bearings almost unconsciously, just as a horse or an Indian does. A dozen things try to tell him. The sun, of course, or a lighter place in the clouds where the sun should be. The moon. The North Star. The shadows and the time of day. The timber types. Moss on the trees. The wind direction. The way the land tilts. Direction of flow of streams. I've known persons to swear by all that's holy that a stream was running uphill. They were so turned around they couldn't believe the running water. Landmarks, as a certain mountain peak. Even sounds such as a train whistle. I said almost unconsciously. Actually a woodsman keeps track of one or more of these things with one small part of his brain while doing something else. The less experience one has, the larger the chunk of his attention that must be given to keeping his bearing by consciously observing the guides mentioned. In new country it's a good scheme after passing a landmark to look at it from the other side. Indians do this, so that if they want to come back that way they will know how it looks.

A compass will tell which way is north. But it won't tell you where camp is. Don't hold your gun close to the compass while reading it. A compass needle is magnetized and a big chunk of metal such as a gun or a car *may* swing it off course. A map is often of little use unless one has a way of orientating it—either by compass, sun, or recognizable features on the map. But with a compass and a map and some careful study one never need be lost.

The Blue Mountain District on the Whitman National Forest had quite a siege of lightning starts one day. Ranger Harry Wolfe had all the fires manned and was organizing follow-ups as best he could. I went out to a fire near the ridge north of Bridge Creek. Five young fellows from the slash disposal crew had been sent to it. They had the fire corralled, but it wasn't mopped up yet. So we shoveled and chopped and stirred and by 10:00 P.M. it was out. We had only one flashlight and I gave it to one of the boys to lead the way to the road and I brought up the rear. Shortly the leader asked to be relieved of the flashlight job, so I took the lead. It was dark. The mountainside was steep, with no trail, and we did some zigging and zagging to avoid windfalls and rock outcrops. Finally we came out onto the road clearing within twenty feet of my parked car. Those boys from a Midwest college thought it uncanny. They couldn't quit talking about it. Actually it was just an everyday case of keeping one's bearings.

The common advice is "go downhill, if you are lost." But there are places you would travel downhill a hundred miles before you would reach a settlement. It may be much closer uphill to a camp or a telephone, or a good trail with signboards. Here's where your map comes in handy. The toughest traveling is invariably along a stream. The brush is more dense, the timber is bigger and thicker, and there are more windfalls. The watercourse may have cut through rock barriers and on the creek one will have to climb steep, high cliffs frequently. Bottoms of canyons are darker at night. The easiest traveling is on a ridge, or a south slope close to a ridge. Less brush, and the timber is more open.

During World War II a group of fliers, in training in Idaho, were forced down far back on the Salmon River. They glided safely to a frozen lake surface. They could not fix the plane. Someone had built a cabin by the lake. No one came in the next two days, so they remembered the advice to follow water downhill. The entire group did that almost to their deaths due to frozen feet. The frozen lake was perfect for putting out SOS signs and rescue would have followed. The same year a seven-year-old girl was lost in a thick forest. She found a cabin at the edge of a clearing and stayed there, putting out a HELP sign made of fir boughs. She did not suffer unduly. The little girl wasn't smarter than the fliers, but she was accustomed to a forested land—they were from the prairies.

If you find yourself without a compass it may be possible in daylight to improvise. Using your watch, point the hour hand at the sun and south will be halfway between the hour hand and twelve o'clock.

One can lose his bearings and get really lost but that doesn't need to produce panic. Before going farther sit down—try to relax. Think how you got there. What threw you off? Study your map. Check direction with your compass. Believe what it tells you. Try to throw off confusion. Take it easy. Rest. Finally, decide what you are going to do. Don't strike off in one direction only to decide you're wrong and reverse yourself. And make sure you don't travel in a circle.

Avoid Night Travel

That's a good rule. Especially if you are confused in your directions. It will be so much easier to get straightened out in daylight, to say nothing about the greater ease of travel. Only a few years ago a hunter west of Baker was apparently finding his way out to the valley at night and he stepped into space from a high cliff. He was found next spring. Last winter (1965) an elk hunter in Union County died that way.

Firemen (smokechasers) on the forests are expected to go to fires night or day. But they have had training in night travel, they have been over the trails on their unit and know their country. They start fresh from a known point, not tired out after a day's hunting. Of course, if it gets dark when you are almost to camp and you know where it is and how you can get there, it would be silly to quit and sleep out. But if you are not sure of yourself, then:

Look for a Place To Spend the Night

Don't wait until dark. Try to find shelter from the wind. That is more important than something overhead. A big rock or cliff is fine. Or a big tree may serve both as a windbreak and overhead shelter. Spruce trees are likely to fill the bill better than pine or tamarack. A real large windfall does quite well. Sometimes a cozy shelter can be found in a hol-

Courtesy Div. Blister Rust Control, Bur. Plant Industry, U.S.D.A.

ON THE CREEK BOTTOMS ONE FINDS MORE BRUSH, DENSER SHADE, AND THE WINDFALLS MAKE THE GOING SLOWER

low, burned in the butt of a standing tree. Boughs can be
cut and used as additional shelter or to make a lean-to. They
can serve also to keep you out of mud or snow. Plan a place
for your fire. It should not be against the rock or cliff or tree
but out far enough to allow you comfortable space between
the fire and backstop. This way your back gets some warmth
as well as your front.

Jackman hunted a few times with his brother-in-law in
Montana who was a professional cougar hunter, Jack Hutchens. When on a trail, he stayed on it with his hounds. When
night came he kicked the snow back from an area seven feet
long and built a fire, usually in front of a large tree or rock.
As he put on more wood, he always put it on the back of the
fire, so, as the night wore on, he had dry warm ground to
sit on even though the snow might be several feet deep. It
takes a small fire, with no coals left behind. The scheme
works, once you find how to do it. One of the chief things
is to avoid slight depressions. Water from the melting snow
all around runs into such places and you awaken wet and
dismayed, in a miniature lake.

One Christmas vacation Norman Henchel and I spent two
weeks in logging camps in the pinery up north of Cloquet,
Minnesota. We finished gathering white pine logging cost
data and a study of logging operations and were headed home.
We planned to catch the evening log train that had no regular
time schedule. There was no depot and the closest logging
camp was a mile from the track. So we went over to the track
before dark to flag the engineer. It was logged-over country
and we had no trouble building a fire. We had no axe and the
snow was well above our knees. The fire gave us light and we
kept moving, dragging in more fuel. It was cold but the heat
from the fire and the exercise of fuel gathering made us
fairly comfortable. Well past midnight the delayed train
showed up and we entered the caboose and went right to
sleep in the heat from the potbellied stove. The next morning
in Cloquet we learned the thermometer had stood at fifty-
two degrees below zero during the night.

Building Fires

Boy Scouts can make fires by friction but it is so much easier to use dry matches. We repeat—never get caught out in the woods without matches! A few slivers of fat wood (pitch) could well be included in the knapsack. Jack says he always used to carry a piece of birch bark with him. Do you know birch bark right off a green living tree makes the best kindling in the world? It contains an oil that doesn't soak up moisture and burns like coal oil. Just make a vertical cut with a knife and peel off a horizontal strip. Pitch wood from a pine stump is tops also, and it can be found most anywhere. Fuzz sticks can be whittled from dry pine or other wood. (Shavings cut with one end left on the stick.) A few of these placed together are fine for starting fires. A short piece

Courtesy Carl Holman Studio, Baker, Ore.
ONE SITE FOR A BIVOUAC
An uprooted tree provides a windbreak and a reflector for the heat from your fire. It may even protect you from rain or snow. Charles Simpson is pictured.

of tallow candle is OK. Dry needles—especially pine—are good. Dry ones can often be found on dead branches or in protected places. Fine twigs broken from dead branches are usually dry enough to ignite readily. Shredded inner bark is often dry and is very good. Wood cut from the central part of a large limb and split into fine slivers is usable.

A person should never want for fuel in the forest once a small fire is started. In wet weather or in winter, standing trees make better firewood than do windfalls. Split sticks ignite more easily than round pieces. Small pieces burn better than large ones. Pieces laid together parallel do better than jackstraws. Both live birch and green Douglas fir wood burn readily when thrown on a hot fire. Dead trees are usually available and are preferable. Lower limbs of large live pines are often completely dead and dry.

Distress Signals

Rifle shots would be thought of first. The nationwide SOS call is three signals of any kind, either audible or visible. The standard distress signal is needed so seldom, it's well to agree among members of a party what the standard signal is and that it will be used. Repeat at regular intervals. A rifle report can be heard a couple of miles if conditions are right. The main drawback is that one can run out of ammunition long before he is ready to quit firing. Save some shells for later use when you know someone is out looking for you. If you have to stay out long, some shells may be handy for shooting animals or birds for food.

At night a good blazing warming and signal fire can be seen for ten miles or more if visibility is good and ridges or mountains do not intervene.

In the daytime, smoke can be seen half a county away on a clear day. A good volume of smoke can be produced by making a lively hot fire, then adding damp bark, or green grass, or even green boughs of trees. Boughs soon curl and scorch and blaze, but in the process they put out a lot of smoke.

The smudge should not be in dense or tall timber or it might not show up. Preferably it should be on high ground, on a ridgetop or on an open sidehill. Sometimes there is a flat rock on top of a cliff. The fire should be in an area visible from some lookout or other occupied station, or in sight from a settled valley. If not in an exposed site, or if it is not feasible to put a fire where it can be detected easily—pile on the fuel and smudge-maker, the bigger the better, and rest assured help will come.

Your real dangers are: prolonged snow; a three-day blizzard; fog. In either case try to hole up and improvise as much shelter as you and inspired ideas will allow. Panic in any one of those three cases is often fatal. In one Oregon case a well-schooled huntsman stumbled into a cave. He crawled in as far as he could go and, exhausted, soon went to sleep. He awakened suspiciously warm and found himself cozily snuggled up to a nice fat mama bear, hibernating.

To be spotted from the air use smoke and supplement it with an SOS signal on the ground. The call for help should be out in an opening, a meadow or clearing. In summer a white material, flour or white cloth strips, should be used, preferably on the bare ground. Make it large. In winter, with snow on the ground, the letters can be tramped in the snow and some dark material added to make the marks stand out. Charcoal, earth, bark, needles, or rotten wood would do.

If a low-flying plane or helicopter is moving over the spot, a person in motion, out in an opening, and waving a flag of some contrasting color, will gain attention, but a fir-bough flag will do. One's coat or shirt, or anything waving, is easier to spot from the air than the arms and hands. Yellow is said to catch the eye more quickly than any other color against most backgrounds.

Emergency Food

First, take a normal lunch along for noon that day. Beyond that it's anybody's choice. You may not need it. You don't want to be packing several days' food each day in the woods, just to have in case of emergency. Yet for the occasional or rare emergency it's fine to have something. In other days a chunk of jerky went a long way. You can chew and chew. It tastes good and provides a lot of power for its weight. Chipped or dried beef is a fair substitute. Raisins give a quick pickup and allay hunger. Many like chocolate bars, but in hot weather they are messy in the knapsack and easily wasted. Hardtack or ordinary crackers keep up one's strength. Dried prunes are good to chew on and the pits, carried in the mouth, keep it from getting dry. Some real hot tea gives me more of a lift and makes the world look brighter than anything. It's great if you are cold or wet and, pound for pound, it can lick its weight in any food I know of. You do need a container to heat the water.

After all starvation won't get a person if he has nothing to eat for several days. Hunger is largely psychological. Jackman has told elsewhere of the old Montana trapper who gave him the best advice he ever had. "Son, don't never git tired till you git to where you can rest and don't never git hungry till you git to where you can eat." If one keeps busy and doesn't dwell on the need for food he can get by. But something nonperishable, lightweight, and appealing to the person involved is worth carrying for an emergency. It helps keep the thought of hunger under control, and keeps the tapeworm from lashing his tail too vigorously.

The cook at a forest camp fixed a lunch for me one morning and as I tied the sack on my saddle he said, "There's a ham bone in there. A man will never starve to death as long as he has a ham bone." After traveling all forenoon I stopped and unsaddled for lunch. Sure enough, the old devil had put in a good-sized ham bone plus a normal lunch and my horse had packed it for twenty miles. But it wasn't wasted. A stray dog took up with me that morning and was with me still at noon.

First Aid

Every camp should have a first-aid kit. Kind and size are of less importance than to know how to use it—or, for that matter, what to do with or without a kit. With Red Cross courses given in every town, every mountain party should have at least one person who has taken the course.

Knowing how to stop bleeding can save lives in the mountains or on the highway. What are the pressure points? When and how should a tourniquet be applied? How would bandages be used? At a Boy Scout camp, training in first aid was being given. Some of the boys were posted out on a

Courtesy Carl Holman Studio, Baker, Ore.
NOT A SUITE AT THE PALACE HOTEL
But you won't freeze to death if you keep a good fire going. The important thing is a windbreak and plenty of firewood.

hillside as patients and first-aid teams were to find the patients and treat them. One of the patients was to be a bleeding case. He waited and waited for his rescuers to come. Finally he wrote a note and left. The lifesavers read, "I bled to death and went home!"

Heart attacks cause as many deaths among hunters as all accidents combined. Change to higher altitude, excitement, and overexertion are too much for some people. The ages given in newspaper reports indicate many people do not know their limitations. If they won't stay home or stay in camp, they at least should take it very easy. Self-restraint is safer than first aid.

Shock, if serious, must have immediate attention. It is that collapse that comes with every serious accident and even with many little injuries. You feel faint, your face gets pale, your skin is moist and clammy, your pulse is weak and rapid. Your mind is dull and you may go on to unconsciousness. There may be nausea and vomiting. Shock may be so severe as to cause death in cases of severe pain or much loss of blood.

Asphyxiation or stopped breathing. Lack of oxygen. The stopping of breathing may be caused by being under water, by electric shock, by deadly gases, strangulation, or by being smothered as in a landslide or snow avalanche. Prompt artificial respiration is necessary or death comes soon. Mouth to mouth resuscitation is the present recommended method.

A partial list of other disabilities includes: food poisoning, sunstroke, frostbite, fractures, burns, wounds, snakebite, insect bites, sprains, strains, or plant poisoning.

The person in camp who knows what to do and does it promptly is a handy guy to have around. You can forgive some mighty poor pancakes and some disgustingly bad shooting from a man who saves your life.

CHAPTER EIGHT

ON THE SUMMER RANGE

Use of Forage To Continue

WHEN CONGRESS AUTHORIZED THE ESTABLISHMENT OF NA-
tional forests, it specifically said "for the production of timber and protection of watersheds." James A. Wilson, then
Secretary of Agriculture, in his first Regulation stated that
national forests were to be so managed and so used that they
would "render the most service, to the most people, in the
long run." He mentioned wood, water, and forage. So multiple use was born, even though the phrase was coined later.
Wilson's pronouncement has been the guide and touchstone
to all national foresters for sixty years. It has helped thousands of users of the forests. It has hurt a few.

From the first, two principles have been recognized. Forage grows in many places on the forests and can only be used
by grazing animals. Range use is important in rounding out
a livestock operation. Most livestock ranchers have four
requirements: winter feed, spring range, summer range, and
fall pasturage. National forest ranges are adapted to summer
use. That's where the calves make their fastest and cheapest
gains; where the lambs develop the bloom that the packers
look for. Should the summer range not be available it would
throw the whole livestock operation out of kilter. Land
needed for hay production would have to be pastured. If
spring and fall range were to be used in summer, more feeding would be necessary. It would mean fewer livestock with

less finish; lower prices per head for the grower; higher prices per pound for consumers.

Now and then we hear a stockman say that the plan is to remove all livestock from the national forests. In many contacts with Uncle Sam's foresters at all levels, from the chiefs down, I have *never* heard that idea expressed. It will never come about because of the wish of the national forest administrators. I seriously doubt if it will ever come about because of the demand from sportsmen and recreationists. The kids will go for hamburgers but most of us enjoy a thick T-bone steak or a tender lamb chop.

Some of the national organizations imply that stockmen have had to fight the Forest Service for everything. Wilson's "most benefits—most people—long run" pronouncement has caused treading on some big toes and has caused some severe temporary pains. By and large foresters and stockmen have been good friends. The Forest Service has had many ex-stockmen in its key positions. Will Barnes, an ex-cowman from the Southwest, was in charge of grazing in the Washington office for many years. Many were on Regional Foresters' staffs and on the forests. They have largely given way now to a new generation of Range Management graduates from the colleges. Many of these boys came from farms or ranches and know what it means to "shovel" hay and "mark" lambs. They know the smell of the soil as well as its chemical makeup.

The goal has always been to act in the best interests of the forage resource, which means also in the best interests of the stockmen. Some adjustments have been made, too, because of the other uses. The majority of stockmen realize they have gotten along exceedingly well at the hands of us bureaucrats. Differences have always existed between renters and land managers. Fences, water, time of turn-in and takeout, other users, special use areas—all such things offer chances for a difference of opinion. It is pretty hard now to find a stock owner who wouldn't agree that creation of the United States Forest Service had been a good thing.

Courtesy U.S. Forest Service

RANCHERS OF ROSEBUD COUNTY, MONTANA, CUTTING CATTLE, WORKING THEIR OWN STOCK OUT OF THE HERD

Neighbors usually join forces for this job

Permits

Prior to the creation (that's a misnomer) of a national
forest, the public domain was truly public. Anyone could
cut timber, graze sheep, graze cattle, run horses. The cattlemen fought the sheepmen and the sheepmen fought among
themselves. It was a case of who got there first or who had
the fastest gun. Sometimes it was a case of eyeball to eyeball. Sheep were run over cliffs or were shot to death. Dead-
lines were established. One of these was known as Deadline
Ridge, east of Rogerson, Idaho, on the Minidoka Forest. A
long smooth ridge running north and south for miles separated the choice cattle range to the west from the more
rugged and more timbered areas to the east. It meant death
to the sheep, and to the herder, too, if a band made an appearance along the ridgetop.

Following the establishment of the national forests, the
first job was to determine the range users. Applications
were taken from all who claimed a grazing interest. Prior
use had to be proven. A nearby ranch owner was given
priority over a nonowner or transient coming from a distance. The following year no stock would be allowed on
the forest without a grazing permit. The permittees had
to be named and the allowable number of stock determined.

Applications were approved and permits issued. A card
record of each permittee was started and a case folder set
up. Two filing designations were used, one for sheep and
goats (S and G) and one for cattle and horses (C and H),
and separate filing compartments were used. I never issued
any permits for goats but they were common in the Southwest and the designation persisted. At Monpelier, Idaho,
headquarters of the Caribou Forest, George Henderson and
I were newly assigned and arrived on almost the identical
day in the spring of 1917. I went out to get acquainted
with the districts and George reviewed the files. A man
named Herb was the lone clerk. As George finished with each
folder he put it in Herb's basket for refiling, as was good

procedure. Herb was rebellious. For some reason he shortly resigned. A while after Herb left we found George's personnel folder filed correctly in the H's but in the sheep and goat file, instead of the personnel drawer. It could have been an accident.

Permits were first temporary and, after three years' use, were on a permanent or preference basis. At first only annual permits were issued but eventually ten-year term permits were the rule, where preferences were established through use and the range seemed adequate.

For years we had far more applications for permits than the ranges could carry. New people, old settlers who did not get permits through prior use, hired help, sons and daughters of permittees would make application. The number outgrew our clerical ability to write individual letters. On some forests a form letter was mimeographed and reasons for full or partial disapproval were listed with a square for

Photo by Charles Simpson, Baker, Ore.
MILLER & LUX CAVVY IN CORRAL
The cowboy is practicing roping with his left hand. He lost his right in the "bight of the line" when he roped a big steer with his lariat fast to his horn. Miller & Lux once owned land almost from Mexico to Washington State. They were probably the largest owners in America.

checking. Reasons might be: no prior use, a minor, a married woman, no cattle owned, no base property, no hay production, own an excess number of stock, a nonresident, an alien. The recipients, often without paying attention to the reasons checked, would write in with some heat that they were not married women, not aliens, or didn't fit another classification. A personal letter of disapproval raised fewer hackles and removed the need to clear up misunderstandings.

One applicant was the "Wheelbarrow" woman of the Snake River. Where she came from no one knew but she first appeared on the scene pushing a wheelbarrow loaded with her earthly possessions. She was traveling north on the Oregon side of the Snake. She found a prospector sick in bed and took care of him until he was well, then continued to make that her home. After several years the prospector died and the Wheelbarrow Woman became owner of the acreage on the river bar. She later decided to sell and applied to the Forest Service for a grazing permit for one hundred head of cattle. She candidly explained that she could get more for the claim if it carried a grazing permit. One lone cow would have eaten more grass and hay than the tract supported.

Limits

What was known as the "protective" limit was determined for a forest or group of forests having similar conditions. Seventy-five head of cattle was the common figure in many areas of moderate-sized ranches. In some diversified farming Utah communities with many permittees with four to ten head, the protective limit was correspondingly small. In areas where large outfits were the rule, a much higher limit was set. In the case of sheep, a protective limit of a thousand was often established because sheep have to be handled in bands with a herder in attendance plus a full- or part-time camp tender, and it was not practicable to run them in smaller bands.

The protective limit worked to the advantage of the "little man" (we had them even before the New Deal days). If there was any surplus range to be allotted, it went first to the man owning less than the protective limit number. He could conceivably build up to the protective limit. Unallotted range sometimes became available in those early years. Some of it resulted from improved use, while some slack was gained when transients dropped out or when non-owners of ranch property were dropped who had been using the range under temporary permits. The owner of less than the protective number was exempt from reductions, providing of course he met other requirements. In short he was the Class A man.

Shortly another limit was established known as the "maximum" limit. It did not mean that the original permittees

Courtesy U.S. Forest Service
CATTLE SHADED UP IN THE HEAT OF THE DAY ON A FORESTED
SUMMER RANGE
On the Kootenai National Forest in Montana

who were granted permits on the basis of prior use could not continue to hold them even if above the maximum. It developed as a result of transfers (touched on as the next topic) and was initiated because of a concern that monopolies of the range might occur.

With a protective limit of seventy-five head of cattle the maximum would run around four hundred head. After the first few years of national forest administration with its shaping up and shaking down, there was an extended period when very few reductions were made. But if reductions were to be made, the permittees above the maximum were vulnerable.

Transfer of Permits

From the start it was held that grazing permits were a privilege and not a right. They could not be sold but could be transferred with the change of ownership of ranch property and a waiver of grazing privilege by the seller. Later it became approved practice to transfer the permit with either ranch property or permitted stock. Livestock taking with them a home on the summer range were worth more than the same stock in the open market. The permits were never knowingly transferred to a buyer of the grazing privilege alone. It had to be with the transfer of permitted livestock or base ranch property.

A reduction in number, usually 10 percent, became common in the case of transfers of permits of owners above the protection limit. A reduction in the case of a transferee was usually not so serious as with the original owner. Possibly it meant that the seller did not receive as much bonus as he otherwise would have, but at least his own operation was not curtailed. Slack gained through transfer reductions was reassigned to qualified small owners. Later such cuts were made only where the range was overstocked. The small owners whose numbers were built up treated this with whoops of delight but the larger owners bucked the idea.

Grazing Fees

Bona fide homesteaders and settlers were allowed to graze up to ten head of milk stock or work animals free. All others were required to pay a fee. The first seasonal rate of five cents for sheep and 20 cents for cattle seemed unreasonably small to people in the Midwest and East. Small increases were made from time to time. By 1919 the minimum rate was sixty cents per head for cattle. The Forest Service came under severe criticism from Congress with a growing insistence upon raising the grazing fees. An appropriation bill rider proposed an increase of 300 percent. This the Forest Service opposed as did the stockmen. They were familiar with the problems of using the ranges. Owners had losses from predatory animals, poisonous plants, straying, theft, and mountain storms. It was not all a bed of roses. A man sitting in a comfortably heated committee room in Washington, D.C., was far from a sudden mountain blizzard in Montana.

Photo by Charles Simpson, Baker, Ore.

HOME ON THE RANGE

The original mobile home. It even moved with the seasons—to the mountains in summer and to the desert in winter.

The Forest Service then agreed to make a range appraisal Service-wide and to establish new fees to be effective in five years or in 1924. Final base fees were 14½ cents per head per month for cattle and 4½ cents for sheep, but were not put into effect for several years. Later annual adjustments were made, based upon the price of beef and lamb. The percentage up or down from the meat price at the time of establishment of base fees was used to determine the grazing fees for the following year. This was an equitable solution and it tied in with the stockman's ability to pay. Of course there was a one-year lag.

There was some demand to sell the forage on a competitive bid basis as is done with timber. The Forest Service consistently resisted this. It would have wrought havoc to the established, dependent, local rancher and would have been damaging to the range, what with new stock and new owners, neither familiar with it. The system in use gives stability and users have a reasonable interest in the welfare of the range. Manager and user get along now remarkably well and both spend increasing amounts to build up the resource.

Allotments

Early individual allotments were made for sheep. They were laid out as equitably as possible, considering forage, water facilities, and routes of travel. Boundaries were marked and maps furnished to owner and herder. This eliminated the old rush to get there first to keep ahead of other outfits.

Cattle allotments were laid out in large natural units and used by several brands as a community range. They might carry only a few hundred cattle, or in instances up to six or seven thousand. These allotments resulted in the creation of local grazing associations.

Courtesy U.S. Forest Service

BLACK SHEEP ARE MARKERS IN A RANGE BAND

Herders can quickly make a rough check for missing sheep. This part of the band shows one marker. The number of range sheep has declined drastically in recent years.

Stockmen's or Cattle Associations

The association would employ one or more salter-herders who rode the allotment, distributed salt, distributed the cattle, kept the bulls scattered, were alert for cattle thieves, kept an eye out for stock with porcupine quills and for those that were sick or injured. They looked out for water holes, maintenance of drift fences, encroachment of stray stock, maintenance of stores in cabins—all the things that any man has to watch for in riding his own range. Those men kept in touch with the rangers, too.

The association purchased the salt for all stock on the allotment. This arrangement gave better results on the range than hit-or-miss riding by the owners themselves and left them free to handle haying and the many other ranch jobs. One of the salter-herders was encountered by an elderly woman in a party whose car bore Massachusetts license plates. On noticing a rawhide lariat coiled up on the rider's saddle, she inquired what it was used for. Said the rider: "We use that for catching calves and cows, and sometimes horses." "Indeed!" said the curious woman. "And what do you use for bait?"

Each cattle association elected a board of directors and boards and Forest Service worked together. In some cases special rules were adopted by the association and approved by the Regional Forester. The Forest Service helped to get compliance. Some state laws specified that a user of the open range must turn out one bull of beef breed for every twenty-five head of cows of breeding age. Some states said thirty cows. The majority of the members of some associations wanted higher requirements, such as registration, or inspection as to the quality of the animal or even the breed. Compliance was achieved through the use of a special rule.

The Dry Creek Allotment on the Caribou Forest, for example, was carrying about seven thousand cattle and none of the owners was so located that he could open his pasture gates and turn out onto the range. All of the cattle were brought to an entrance point with holding corrals and counting gates. The stockmen had adopted a special rule and elected a bull committee. Each bull had to be registered, must be of Hereford breed, and must pass inspection for type, size, health, and serviceability. If one didn't qualify, he stopped being a bull on the spot.

One user on this allotment was a thorn in the flesh of the other owners. The others ran cows and calves. He had a permit for five hundred head and each spring would ship in by train two hundred and fifty thin yearling steer calves from southern Utah. He got them then for about seventeen dollars a head. They arrived just in time to go onto the forest. He also put back on the range his two hundred and fifty head of two-year-old steers. He had a contract to supply beef

LAMB CHOPS ON THE HOOF

Fat lambs from the Lolo National Forest ready for shipping from Lathrop, Montana. In the middle of the summer lambs are sorted from the ewe bands and those fat enough are sent to market, the others to feed yards.

year-long to a slaughterhouse in the coal mining town of
Kemmerer, Wyoming. By mid-July he would start shipping
a carload of steers periodically, gathering them from the
range. Come fall, he put the rest of the two-year-olds in the
fattening pens and roughed through the yearlings. Under
his system he had no expensive bulls, no dry cows, he had two
years' cheap grass against little more than one expensive wintering bill, he had a sure market and a periodic income. In
addition he had a larger permit than the other association
members. The only legitimate complaint was the disturb-
ance he caused among their cows and calves through riding
and gathering fat steers during the summer and early fall.

In another association the largest owner was a successful
stockman who was born on the Emerald Isle. He would sil-
ently sit through the discussion as to how large an assessment
was needed for salter-herder wages and purchase of salt. As
a vote was about to be taken, he'd say, "Oi don't know why
we have to buy saalt for the ceows. We never ust to put out
any saalt on the range. The ceows can get what they nade
by lickin' the eearth." But they would levy the assessment
and he would pay his bill which was about half the total.
These stockmen, incidentally, pay for salt for thousands
of deer, elk, porcupines, bear, squirrels, and all other salt-
loving wild things. In some places wildlife groups help out
—such folks as Rod and Gun clubs and sportsmen's associations. The arrangement varies by states. In Oregon the
state distributes salt for wildlife by airplane or by ground
means. It is placed where it will attract game away from
livestock ranges and ranch property.

Another association president had a faculty for getting
every motion approved. When he called for the vote he'd
say, "All in favor say aye," giving lots of time for slow votes.
Then he'd say, "All opposed, no—carried," all in one breath,
with no time at all for even the fastest voter to get in a "No."

The salter-herder on the Shoshone Cattle Allotment on the
Minidoka Forest provided an excellent demonstration of distribution of cattle by placement of salt. He put out a supply

of salt on the top of the Deadline Ridge which still was used as the boundary line between sheep and cattle ranges. The cattle found the salt, then elected to go east for water and feed, instead of west onto their home range. A terrible scream went up from the sheep permittee that the cattle were eating him out. And they were.

Trespassing Stock

Handling grazing trespass is an unhappy and thankless job. Few persons like to do police duty. From a range management viewpoint it is most important. If a range or an allotment is capable of supporting a given number of livestock and that number is permitted, any additional stock on the range becomes an overload. They are a serious detriment to the forage stand. They consume feed to which the permitted stock are entitled. No compensation is received by the landowner for such unauthorized use.

Three types of grazing trespass occurred in the early days.

1. Transient bands of sheep frequently cut across forest ranges or took a circle inside the boundaries and were gone before the rangers were aware of it. Scores and scores of bands of sheep wintered on the public domain in Nevada and Utah. They trailed northerly in spring, across the Snake River country, and on up to the Montana line toward Yellowstone Park. In the fall they reversed and headed toward the winter deserts. Water was their controlling factor. A couple of burros carried the camp outfit, grazing right along with the sheep. The U.S. Grazing Service (later to become known as the Bureau of Land Management) put an end to most of the transient sheep grazing.

2. Grazing unpermitted cattle brands or unbranded horses was quite common, too. With cattle this was usually limited to small numbers. A few head of strays would leave an owner's pasture or his private range and get up onto the higher land and better feed on the forest. In some cases a little encouragement was suspected, such as a thump on the cow's

rump with a long riata. But the early rangers were expert at "reading" brands and the permittees didn't approve of an outsider sneaking under the circus tent, so detection was easy. The more difficult job was getting the guilty party to remove the stock. To make a trespass case and charge for the forage plus punitive damage, it was necessary to do more than find the stock. Numbers involved and duration of illegal use had to be proven to make the case stand up in court.

With horses another situation was involved. They can winter out, for they will travel to south slopes and will paw snow to get grass. They eat the range more closely than cattle and do more damage. Not true wild horses, they multiplied and sizable numbers of unbranded horses resulted. In some localities bands of true wild horses utilized the range. The Kaibab National Forest in northern Arizona supported

Courtesy U.S. Forest Service
WATER DEVELOPMENT AS AN AID TO FULL AND PROPER
USE OF FORAGE
Sheep troughs developed for water storage, Beaverhead National Forest, Montana

hundreds and hundreds of wild horses, known as broomtails.
In 1919 I saw a number of bands, but none at close range.
They made a thrilling sight with an outstanding stallion as
the leader, their tails and manes fanned out as they sped over
a ridge.

In Supervisor Dana Parkinson's office in Salt Lake City,
en route to the Kaibab, I was interviewed by a reporter. He
asked questions, including one about the problems on the
forest. In an unguarded moment I mentioned the bands of
wild horses. I never saw the news article but shortly came the
letters. Most writers wanted to know how we were going
to rid the range of wild horses. Would we use rifles or would
we conduct roundups? Some wanted jobs as riflemen, others
had top saddle horses and wanted jobs as riders. Others offered advice or proposed schemes guaranteed to get the job
done. Reporters and other writers wanted pictures. Sentimentalists wanted to be sure the horses would not suffer.

3. The most troublesome was the grazing of stock bearing permitted brands but in excess of the number authorized.
This wasn't a problem in the case of sheep, as bands were

Courtesy U.S. Forest Service

WATER DEVELOPMENT
Pond built to hold snow water and rain for stock use

counted. Of course not all foresters were good sheep count-
ers. Nor was it a problem with cattle brought to the forest
and turned on at a given date and place. In many locations
the cattle were turned out on early range, either public or
private, and as the low range was used up or water became
short, the stock drifted onto the higher and greener forest
ranges. If the owner had more cattle than his permit called
for, they often went right along with the others. If an owner had 100 cows with only 80 permitted, the other 20 had
an awful job learning that they were illegal.

Devices were tried to correct this. Where the cattle were
wintered on feedlots, it was common to make a feed lot
count and ask the stockman to account for his disposition
of the number above his Forest permit. On some allotments
a distinctive paint daub was applied to each owner's cattle
up to his authorized number. In other cases the brushes
of the cows' tails were docked. The most effective was to
furnish distinctive ear tags, each numbered serially. If a
man had a permit for 100 cattle he would be given 100 tags,
or the ranger would insert the tags, the owner putting the
cattle through the chute. The government bought the tags.
Often a distinctive shape was used. One side was the size
of a silver dollar and could be seen for a hundred yards or
even more when the light was right. It made extra work
for the stockmen and it was undesirable to handle the cows
when heavy with calf. Not many were guilty but the in-
nocent suffered along with the culprits.

Chief Forester Greeley went over our regional estimates for
funds one year early in the twenties. At one point he leaned
back and laughed and said, "Where is that wild and woolly
West? Here in District 4 the rangers salt the deer and the
cows wear earrings." The salt was for overpopulated deer
ranges and the burden was too heavy and not right for the
stockmen to carry.

Predatory Animals

The men of the Biological Survey (now known as Fish and Wildlife Service) were primarily responsible for predatory animal work on the forests. Yet the forest rangers lent a hand. They were often first to know of losses and where to locate the trappers. They made cabins available and helped locate old horses and other bait. In the early days they almost always carried a gun and accounted for many a coyote in the course of their travels. I've seen Supervisor McCoy, who was quite a Deadeye, Dick, knock over coyotes from the seat of his white-top buggy with a six-shooter. A man in Morrow County known as "One Shot" could hit a coyote running through the timber at two hundred yards. Orville Cutsforth, of Lexington, asked him if that wasn't a difficult target. He said "No. All targets are the same size. You shoot at the center."

Coyotes were the chief marauders, with killer bears a poor second. Timber wolves, practically extinct in our territory now, were known killers fifty years ago. We had a mounted wolf head hanging in the Oakley, Idaho, forest office. He was entirely white and his fangs were blunted and rounded from years of use. He was credited with killing scores of full-grown cattle before he was finally dispatched.

Clarence Nelson, ranger on the Cassia Division of the Minidoka in the years 1912-14, took his vacation in the late fall and trapped coyotes on his district. William Bedke, an old German cowman, didn't seem to regard forest rangers highly. When he heard of Nelson's avocation, he snorted, "In order to catch de coyòtee, you've gotta know more dan de coyòtee."

Range Improvements

Immediately after the forests were established, a start was made to improve ranges, so as to use the forage better, increase it, and reduce livestock losses. Some work was done by the stockmen alone, some by the government alone, and more with cooperative effort. Sometimes the Forest Service would buy the material and the stockmen would do the work. Appropriations were small but have steadily expanded.

Water development was a favorite. A new stock watering place often saved trailing considerable distance, or made use of forage previously wasted. Many a mudhole or wet spot on a hillside was converted into a good watering place. The spring would be dug out, source of water explored, water concentrated in one spot, and piped out to troughs. Sometimes a concrete box could be used. The stock would be excluded by fencing. An excellent plan was to fill in the spring hole with big rock, then smaller ones, and finally completely cover with earth. In inaccessible areas, logs were hewn or burned out to make troughs. Others were made with plank. A tiny trickle, if conserved, will water a lot of stock.

Courtesy U.S. Forest Service
RANGE FENCES VARY GREATLY. SOME FOLLOW CUSTOM OR HABIT,
AND SOME ARE DICTATED BY CONDITIONS
A figure-four fence built by the CCC's in 1933. Whitman National Forest, Oregon

Poisonous plants caused serious losses. Tall larkspur hits cattle hard. A beautiful royal purple, but stockmen hate it with reason. It grows early in spring and is concentrated in moist basins. I've seen five or six dead cows within a hundred yards of each other and no farther from a nice big patch of larkspur. Hand grubbing is a simple and effective remedy. Poison parsnip or water hemlock is death to cattle, too. It usually grows along ditch banks or in the really wet spots. It is more frequently found on foothill ranches than in the mountains. Lupine, when the pods have formed, is poisonous to sheep. This is particularly true if they are on the trail and hit a patch of lupine when they are hungry. It doesn't seem to bother them if it is encountered when they are filled up or when grazing normally. Lobelia is another plant that causes serious sheep losses. Most deaths I have seen occurred on driveways. Our remedy was to post poison plant areas and reroute driveways to avoid plants. A spray is available now that kills the weed.

Courtesy U.S. Forest Service

RANGE FENCES
Stake-and-rider fence, 1928 vintage. Whitman National Forest, Oregon

Fencing

For years barbed wire was anathema to every cowman. But fences were eventually needed in the management of the forest ranges. First was fencing of forest boundaries to prevent unauthorized stock from trespassing. Boundary fences also served to prevent use of the range before the opening date.

So far as possible allotment boundaries were located so that natural features would prevent mingling of stock from two allotments. Where this was not feasible allotment lines were eventually fenced. This saved riding by the stockmen, avoided losses, and helped achieve the planned use.

Then came zonal fences. Many cattle allotments had both low and high range. The low range was ready first. But cattle, much like wild game, want to work up, "crowd the snowbanks." The stock in their search for green, tender feed, leave mature grass behind and trample the young grass and wet soil. So cross fences became in order, to hold the stock until the later feed was ready.

It was always known that forage plants didn't live forever. Most species produce seed that provide replacements. Others spread by root stocks below ground, or by runners at the surface, with new plants developing short distances from the parent plant. Others use both methods of reproducing. But we were slow to do something about it. If grazing prevents seed maturity or development of tender new plants, or weakens root systems by not permitting enough green leaf surface to cause adequate root storage or development, future production is reduced. Rest for the range is indicated. No allotment can be rested completely in any one year. So rotation of use was worked out. A portion of an allotment might be withheld from use one season until after seed maturity. The next year a different area would be deferred. But to rest the different portions of an allotment, fences had to be built.

This resulted in a series of pastures, with the cattle moved from one to another according to plan. This took into account early and late range as well as resting for seed maturity and vegetative regrowth.

Range experiment stations, both Federal and state, led the way in demonstrating the advantages. In comparatively small areas they established completely controlled use. Number of cow days in each pasture was recorded. Cattle were weighed to determine gains, and marked grass plots were photographed and plants counted. Changes for better or worse were determined over a period of years. Intensity of grazing was studied to determine the relation of rate of stocking to plant vigor, forage production, and gains in weight.

This accurate information was valuable to range manager and livestock owners alike.

Numerous types of fence were designed. The split rail fence common on many pioneer ranches was used but little. The splitting was too much like work. Log fences were more common. Logs eight or ten inches through were laid four or five logs high. An equal number of notched pieces about two feet long were laid at right angles. One log each way would rest on one cross piece. The ends thus made a sort of crib effect.

Stake and rider fences were common. There were two kinds. In one case an average-sized post would have a two-inch hole bored in its upper portion. A smaller piece with a turned end would have this small portion stuck into the bored hole in the post proper. The two pieces at right angles would be set on top of the ground in a leaning position. The other variety had the two pieces spiked together. Three wires were usually staped to the main piece and a single wire attached to the other leg to hold the leg from being moved, and to keep animals from crowding the fence.

The Forest Service usually adopted the figure-four fence. It consisted of an upright post not set in the ground. A horizontal post-sized piece was laid on the ground at right angles to the fence line with ends extending about equal distance. A third piece was spiked to one end of the horizontal piece and to the post about two-thirds of the way above ground. This prevented the post from tipping either way.

Steel posts have come into use. They don't burn and they don't rot. But drifting snow bends them and they aren't as rigid as wooden posts. They can't be driven into rock and transportation to isolated locations may be expensive.

Stockmen have cooperated extensively in building fences on their allotments and even in some instances have borne the entire cost.

Reseeding

Not much was done in the reseeding of range plants until the late thirties.

At first the seeding was confined to areas that had recently

Courtesy U.S. Forest Service

RANGE FENCES
Log fence on a mountain divide. Fishlake National Forest, Utah

burned over. The ashes and exposed mineral soil made good seedbeds and, if seeding is done promptly, the grass plants get a break with weed competition. A stand of grass started promptly does four things. It stabilizes the newly exposed soil, helping to keep it from washing and blowing away; it provides quality forage for domestic livestock and game animals; it tends to prevent "dog-hair" thickets of new tree growth, particularly lodgepole pine, while still permitting enough reproduction of tree species to make a timber stand later; and good stands of grass and legumes keep out weeds and unwanted brush.

A number of forage plants have been found to be satisfactory. Some of them are pubescent wheat grass, meadow foxtail, Kentucky blue, Canada blue, orchard grass, some of the fescues and, where not too wet, Nomad alfalfa. On

Courtesy U.S. Forest Service

RANGE FENCES
Jackleg fence built with lodgepole pine poles, Montana

the Big Cow burn in Oregon in the fall of 1939 a planting of common everyday rye grain was made. In 1940 it produced a thick stand, waist high, and the grain matured although the elevation was 6,800 feet. In 1941 an excellent stand grew thick and rank, this time as a volunteer crop. But the third season not a spear showed up. Being an annual it is not a permanent pasture grass, but it did prove its worth as a quick soil protector.

Where tractor logging is practiced, as much as twenty percent of the ground cover is disturbed, leaving the soil exposed. This occurs on skid trails, temporary or spur roads, log landings, or where concentrated slashings are burned. Here is an opportunity to prevent washing of the soil and to produce forage. The additional sunlight resulting from the logging job provides an ideal condition for grass seedlings. A variety of grasses are suitable. Locally we like orchard grass. It's a fast and early grower, has a good root system and high palatability.

Natural openings often are in need of treatment and can be made more productive than they are. These vary greatly, ranging from wet meadows to dry open hillsides to the poorer soils on exposed plateaus, benches and ridgetops. These are the roughest reseeding jobs. There is a reason why nature is not producing a tree crop on these areas. Too wet, too dry, soil too poor. Yet experiments demonstrate that often good crops of forage can be produced. Crested wheat does well on the drier sites. On each site something will grow.

Often conditions can be improved so that grass seeding is feasible and justified. Contour trenching has been tried. Gophers and other rodents can be eliminated. Unusable weeds such as docks and wild sunflower can be killed by spraying. Wherever sagebrush is found it can be removed with newly developed power machinery and the sage replaced with valuable forage plants. The bigger the sage the better the soil.

Reductions

Reductions in grazing permits, including reductions in number of stock and shortening of the period of grazing, have been more disturbing to the stockmen than anything else about their use of the national forests. It often means running fewer stock or finding other places to keep them, changes in the ranch setup, and a financial loss. Yet the changes were the result of existing or developing conditions and not because of the desire on the part of any of us bureaucrats to work a hardship on the stockmen.

The directors of one of the associations on the Fillmore National Forest in Utah met with Supervisor John Raphael in his office to protest a proposed reduction in numbers on their allotment. Johnnie was quite hard of hearing and he had a habit of cupping his hand to his ear. Discussion was lengthy and at times heated. John reviewed the conditions and insisted reduction was needed to enable the forage plants to hold their own. After hours had passed one of the directors dropped his voice real low and said, "Hell, we're not getting anywhere. Let's go down and have a drink." Johnnie, reaching for his hat, said, "That's what I say."

The Caribou National Forest had been established less than ten years when I was assigned there as supervisor. It was my first supervisory post. I was overwhelmed to find that grazing permits covered 31,500 cattle and horses and 187,000 sheep. Assuming 1,100 head to an average band, that's 170 bands of sheep. Lambs don't count against the permit and aren't paid for. Assuming 100 percent lamb crop, that's crowding a half million sheep and lambs. Every ewe and each lamb has one mouth and one belly to be filled. That takes a lot of grass. I sat and stared at the steam radiator for some time and then it dawned on me that there were five District Rangers on the job and that was the answer.

I'll digress and tell of those five men.

Bob Gordon's father was a judge in Texas. Bob was a school dropout to become a cowboy. He was from good stock and smart, with a fine personality. Days were never too long or weather too tough. He had a wry smile and a slight Texas drawl. Everybody loved him. He knew his job.

Frank Butler was an entirely different sort. Broad-shouldered and heavy-voiced, he had a big, overhanging nose, heavy eyebrows. Tobacco juice colored the corners of his mouth. One glaring look at a recalcitrant Basque and that herder would say, "Yes, sir! Mr. Butler, yes, sir!" He was a local native son and talked the stockman's language. He meant business.

Charlie Spackman was still another type. He was brought up on a horse and got most of his schooling there. He was bowlegged. This was aggravated because a horse had fallen on him when he was a boy and his broken leg hadn't been set

Photo by William Farrell, County Extension Agent, John Day, Ore.
GRASS SEEDING HAS BEEN SUCCESSFUL IN DIFFERENT CIRCUMSTANCES
Result of grass seeding on skid trails and other ground disturbed by logging operations.
Grant County, Oregon.

properly. Spak said he was like the bowlegged cowgirl—he always had trouble keeping his calves together.

One day we were in the front yard going over district problems when a salesman came by and started his spiel.

Courtesy U.S. Forest Service

GRASS SEEDING

Formerly this meadow was bare. It now produces close to a ton of forage per acre.
Logan Valley, Malheur National Forest.

Spackman stopped him and said, "The boss is in the house, I just work here."

Another ranger once said that a forest ranger should either marry a squaw or have two wives. His station was back in the hills and his wife was fearful about staying alone. It wasn't that way with Spak. His wife was self-sufficient, with a large family, some of them nearly grown. Spak could be gone a week or a month and things kept on at an even keel at home. A cattleman asked him once how many children he had. Spak replied, "Well, I haven't been home for ten days but we had eleven when I left." He lived on the range and knew what was going on. Neither cattle nor sheep, nor their attendants operate on an eight-hour shift, nor did Spackman.

Camas Nelson was on the bookish order. He was well read. Methodical and orderly, he kept good records. He

GRASS SEEDING

Grass nursery and experimental area, Coffee Pot Flat, Fremont National Forest, Oregon. The Extension Service, Soil Conservation Service, and the Forest Service cooperated on this and other similar projects. *From left*, Walter Holt, O.E.S.; Merle Lowden, U.S.F.S.; and Waldo Franzen, S.C.S.

didn't ride so many miles in a week as some but he didn't do much backtracking. He got results and without rocking the boat.

James Bruce, the youngest of the five, had been in school the longest. "Reasonable," "responsible," "congenial," describe him. Jim was a willing worker, fitted into the organization, was accepted, and had the approval of the users. Yet nothing was slipped over on him. He worked on the Caribou Forest until he reached retirement.

There was a good team. Each one pulled his load.

The Caribou was an outstanding grazing forest. It was at a relatively high elevation. Almost no part was lower than 6,000 feet and it went to 9,500 feet, with a few points higher. This meant heavy snowfall and ample water. The soil was fertile and extended over ridges with very little rock exposed. The topography was not smooth but it was

Courtesy U.S. Forest Service

GRASS SEEDING

Sagebrush plow developed by the Forest Service to remove brush preparatory to seeding grasses.

not too rugged to be used fully by stock, especially sheep. The northern slopes supported young growth timber, not too dense. A grass sod covered much of the range, with some palatable weeds and browse. Waste land was almost totally absent.

The Caribou was first in the region to be covered by an intensive grazing survey. It was done by trained men. We had, in 1917, a forage-type map on a really intensive scale. It was colored and broken down into very small units. On an overlay were shown the surface acres and forage acres within each small type. A forage acre is defined as a fully stocked acre of 100 percent palatable forage. From previous studies it had been determined how many forage acres were required to support one cow or one sheep for one month. Thus the carrying capacity of any given allotment was determined. This was checked with usage records and the judgment of forest officers and stockmen. Our stocking was in conformity with the range survey figures. We thought we were all set.

Supervisor Rupp has informed me that the carrying capacity in 1964 of the "Old Caribou" (what was part of the Cache Forest is now administered with the former Caribou) stands at 10,630 cattle and 120,677 sheep, and the permits correspond.

That is a reduction of 64 plus percent in the number of cattle and 35 percent in the number of sheep. It has occurred during a period of forty-five years, or over 1 percent reduction per year.

What factors caused the reduction? The Supervisor was asked what percent of the reduction in carrying capacity is attributable to:

1. Encroachment of timber growth.
2. Competition from game animals.
3. Recreation needs.
4. Deterioration of forage growth.
5. Other.

Richard Sanders, of the Caribou office, reports they have no firm statistical data upon which to base percentages. But he said that if they were to place the above items in a sequence as to their overall effect on the grazing resource it would be 1, 4, 2, 3, 5.

He adds, "There has been encroachment of timber onto former open rangeland. The change in forage composition has resulted in a reduced carrying capacity. Deterioration of forage composition and density has also resulted from past overgrazing use by domestic livestock. Competition from game was very slight in the early 1900's but has become a serious problem today."

No one is called upon now to defend the carrying capacities determined in 1916-17, but I believe the determination was accurate and, at that date, the capacity was not overestimated. It was a beautiful range country. The error was in not being able to look ahead and see what changes might occur. Sanders puts timber encroachment as the Number 1 cause of reduced carrying capacity. We should have anticipated that. With fire controlled, it is inevitable that where seed trees are present, new seedlings will spring up. It's also inevitable that little trees will grow into big trees and a big bushy tree will cast a bigger shade and appropriate more water. Under timber shade pine grass, an almost worthless plant, takes over. Valuable forage plants are displaced.

This is an insidious thing. Those bellies have to be filled. If the good grass under the evergreens disappears, the cows or sheep nip a little more often or a little closer somewhere else. Some plants that would have produced seed don't do it. Some with low vitality die. Changes are gradual, unnoticed. The cows come home in good flesh. The forage looks all right. Sooner or later someone wakes up. Whoa! this allotment has too many stock, the range is deteriorating!

The little evergreens aren't the only culprits. Suppose you have a stringer meadow along a creek. There is good grass feed. It's close to water. There is shade close at hand.

The cows like it. They stay there, especially if some salt gets dumped there by mistake. The grass doesn't go to seed. It gets trampled and is kept from rejuvenating itself through ample green leafage. The carrying capacity shrinks.

Or a sheepherder has a favorite campsite where he likes to camp. The sheep go back and forth over the same ground too often and lo! the forage doesn't last as long as it once did. Maybe another area up on top is easy to herd; the sheep like it there. It gets more than its share of use. When feed on these spots isn't there any more, the sheep, to fill up, hit another area harder than they should. Range deterioriation is cumulative.

Many reductions of tomorrow can be headed off by range improvements and good management today. Those from timber encroachment and game competition we may have to live with.

In 1919, at Kanab, Utah, I saw a horrible example of the results of severe overuse. It was Sunday and I was at the home of Supervisor John Roak of the Kaibab Forest. We noticed a rumble like distant thunder. Jack said, "It's a flood. Let's see it." We dashed across the village to the creek.

Here was a chasm 35 feet deep and 150 feet across, with raw vertical sides. In the bottom was a moving mass, half mud, half water, almost too thick to flow. Now and then came a muffled sound and a splash as a section of sidewall, undermined by the current, would slough down and join the moving mass at the bottom. A miniature falls a couple of feet high would slowly move upstream as the bottom of the channel melted away and it, too, joined the movement downstream. Thousands and thousands of tons of earth were moving to the Colorado River.

This was Kanab Creek, meaning "willow" in Indian. Pioneers said that when they first came to Kanab they could ride a horse across Kanab Creek anywhere.

Mormon pioneers settled Kanab and built their homes in the village and farmed or raised stock nearby. They had flocks of sheep and goats that were herded on the hilly range in the morning and brought to the village edge at night. This was repeated day after day and year after year until the grass and sagebrush and all vegetation on the hills was eaten out. The country is subject to severe localized cloudbursts of short duration. When one of these storms hit those eaten-out ranges, the water ran down as on a roof, picking up soil as it went. The soil had no absorptive power and no vegetation checked the runoff.

A complete range betterment plan for any forest will entail a whole book. We have made a dash at it in this chapter. Every mountain meadow is different from a scabland hilltop with only six inches of soil.

Forests differ. In the Blue Mountains of Oregon, shoestring meadows usually make up about 15 percent of the area, but

Courtesy U.S. Forest Service
TREE GROWTH READILY INVADES RANGE TYPES
In time this competition reduces the amount and quality of the forage crop. Montana area.

supply roughly 85 percent of the forage. An increase of deer and, in the early 1900's unrestricted use by livestock, tended to strip the natural meadows of grass. Whereupon the watercourse down the center cut deeper and deeper, thereby draining the sides. These beautiful meadows then became nearly desert in type, or else filled in with thicket-type trees. In either case the grazing for both wild and tame animals was reduced with little beneficial increase of timber. This is a tough problem that can be cured if started early enough.

CHAPTER NINE

FUN ON THE FORESTS

Available Information

NATIONAL FORESTS ARE FREELY OPEN FOR RECREATIONAL use by the public. One or more national forests are easily accessible to nearly every community in the West. Most families can visit the forests several times each season. The wide range of recreational uses encourages frequent visits. More forest roads add to the ease of travel and make more features available. Until 1964 recreational use of the forests was free except for a small charge for certain improved campgrounds on an experimental basis. Beginning in 1965, a seven-dollar license allows entry during the year to all national parks, the use of all national forest campgrounds, and access to any Federal recreational areas. The income is earmarked for recreational development.

Printed Material Available

National Forests—Recreation Folder—including map (each Forest)

Wilderness Areas—Map and Folder (each Wilderness Area)

Regional:
    National Forest Recreation in Idaho
    National Forests of the Pacific Northwest
    Multiple Use Highlights—Pacific Northwest
    National Forests for People—Intermountain Region
    Northern Region National Forests

    Directory—Public Camp and Picnic Areas, Northern
      Region
    Ski—Northwest

National:

    Skiing—The National Forests—America's Playgrounds
      (including directory of ski areas)
    Trees of the Forest—Their Beauty and Use
    Campfire safety
    Camping—The National Forests
    What To Do—When Lost in the Woods

And many more.

Such information may be picked up at one of the many
ranger stations or at a Forest Supervisor's office, or it may
be obtained by writing to one of the Regional Foresters or
to the Chief Forester's office in Washington.*

---

* See Appendix for complete directory of forest offices.

Courtesy U.S. Forest Service
THE FOUR MURPHY BOYS ON THE TRAIL AT CHINESE WALL
Here the Bob Marshall Wilderness Area in Montana straddles the Continental Divide

Wilderness Areas

On September 3, 1964, President Johnson signed the Wilderness Act. It established the National Wilderness Preservation System, at present composed of nine and one-tenth million acres. All of this had been previously classified as wilderness areas in the national forests. Under the Act the U.S. Forest Service continues to administer its wilderness areas. Grazing use may be continued. Mining claims may be staked until 1983. The Act sets up a ten-year study period for consideration of adding primitive areas, roadless areas on the forests or national parks, Wildlife Refuges and Game Refuges.

The wilderness areas range in size from half a million to nearly two million acres. Most include rugged, high-moun-

Photo by Oregon State Highway, Travel Division
SOUTH SISTER MOUNTAIN IN THE THREE SISTERS WILDERNESS AREA
Here is the largest of the Green Lakes. It can be reached only by foot or on horseback

Photo by Bob Bailey, Enterprise, Ore.
EAGLE CAP MOUNTAIN IN THE EAGLE CAP WILDERNESS AREA
Mirror Lake is one of many in the Lakes Basin. Whitman National Forest, Oregon

Courtesy U.S. Forest Service

THREE SISTERS MOUNTAIN IN THE CASCADE RANGE

Area beyond Scott Lake is in the Three Sisters Wilderness Area. Willamette National Forest, Oregon

tain country, with sparkling blue lakes and tumbling streams. Some are nonmountainous, as the Minnesota-Canadian Wilderness north of Lake Superior. This is a vast roadless area dotted with hundreds of lakes of all sizes offering endless variety for canoeing. In 1911 another fire guard and I canoed on one of its districts that was under fire protection for the first time. During the whole summer we saw only two persons on the district, a photographer for the Duluth Iron Range Railroad and his Indian guide.

These priceless solitudes are to remain undisturbed by road-building machinery or by any mechanized vehicle. They are to be visited only on foot or with horses, or by canoe in the lakes country. They will continue as portions of once greater wildernesses, for hardy members of future generations to appreciate. Such persons will still be able to explore these last frontiers, travel little-used trails and observe the vegetation and animal life much as it was in the days of Lewis and Clark.

Most of the designated areas is in the Alpine type or close to or above timberline. Yet enough timber is included to preserve that type of cover in its virgin condition. Proposals for new or additional wilderness classifications may induce bitter controversy and if so, most of it will be concerned with those areas bearing commercially valuable timber. Decisions should be made on the basis of relative need.

Trail Riders of the Wilderness

Beginning in 1933 The American Forestry Association organized the Trail Riders of the Wilderness, a nonprofit organization. Its purpose is to provide experienced and organized leadership so that any man or woman with average outdoor experience may explore and enjoy true wilderness areas lying beyond all roads, uninhabited and unchanged. The United States Forest Service and the National Park Service cooperate with the American Forestry Association.

For a reasonable fee the Trail Riders of the Wilderness in

1965 provided eighteen wilderness tours, into fourteen different areas. Among them are Bob Marshall Wilderness in Montana; Selway-Bitterroot Wilderness, Montana and Idaho; Teton Wilderness in Wyoming, and Glacier Peak-Lake Chelan in Washington.

The organization employs expert guides, packers, and cooks. Riding horses and pack animals trained for mountain travel are supplied. A doctor is assigned to each expedition. Foresters or park rangers usually ride with the parties and the American Forestry Association has a representative on each trip. Each party is limited to twenty-five riders (less in a few cases). The trips cover from nine to eleven days. Nine to fifteen miles is the usual day's ride, with a new camp each night except for one or two stopover camps for rests or side trips or fishing. The riders furnish only sleeping bags, clothing, and personal items.

Courtesy U.S. Forest Service
TRAIL RIDERS PACK TRAIN ON DANAHER CREEK TRAIL,
FLATHEAD NATIONAL FOREST, MONTANA
The pack string moves camp and all equipment and groceries to the next night's campsite

Many are so enthusiastic after one of these wilderness rides into a remote mountain fastness that they repeat with one or more rides to other wilderness areas.

Riding and Packing

The number of saddle horses is higher now, in 1965, than it was fifteen years ago. Sheriff's posses, riding clubs, and youth riding groups are common. Horses are still used on livestock ranches, while resorts and dude ranches have saddle and pack stock. The result is that more people are taking horseback trips into the forests. With horse trailers, people in most of the Western communities are within easy distance of one or more of the national forests and this is one

Courtesy E. R. Jackman, Corvallis, Ore.
AWAY FROM IT ALL IN THE WALLOWA MOUNTAINS,
WHITMAN NATIONAL FOREST, OREGON
This is the Eagle Cap Wilderness Area. Those who go there go on horseback or afoot

of the popular recreations. This may be just a means of transportation or it may be the sole purpose of the trip.

Ranger Gene Wilmoth and I had an odd experience one summer while looking over the Catherine Creek cattle allotment in eastern Oregon. As we rode out onto Dead Horse Flat from the south we saw a lone horseman top out onto the flat from the north some four hundred yards away. Shortly another horseman appeared and then another and another until sixteen had come into sight. It reminded us for all the world of a shooting gallery where the moving targets pop up into sight and move across the board. "The Mustangers," a ladies' riding club from La Grande, Oregon, were enroute to the Eagle Cap Wilderness Area. We saw them again that evening as they camped at Catherine Mead-

Courtesy U.S. Forest Service
WOMEN RIDERS WITH A PACK OUTFIT. BITTERROOT
NATIONAL FOREST, MONTANA
On the Bitterroot Trail

ows, hobbling the horses and turning them out to graze. Some real horsewomen were among them. Only groceries and kitchen equipment rated pack stock, the bedrolls and full saddlebags went with the riders, cavalry style.

Photo by Photo-Art Commercial Studios, Portland, Ore.
TO THE PAVEMENT-BOUND, THIS IS A NEW AND A BETTER WORLD
Mountain air, grand views, and the smell of the woods

Mountain Driving

Driving the mountain roads just to have a change of scenery is popular with an increasing number. There are overlook points where motorists may park and enjoy the view. Some roads overlook fine mountain streams; others may go through attractive stands of timber, or through selectively cut stands where the results of improved forestry practices are apparent. Signs often tell the story, giving facts about the operation or the forest. Because of the miles of country covered, many game animals and birds and other wildlife may be seen especially toward evening and when the wild things are not so wary as during hunting season. Many will appreciate the wide variety of wild flowers and plants.

Roads now lead to many fire lookout stations from which the visitors can see miles in all directions. A good example

Photo by Oregon State Highway, Travel Division
WINTER SCENERY (OREGON) HAS ITS OWN SPECIAL APPEAL

is Hat Point in Wallowa County, Oregon, where Hell's Canyon of Snake River is visible 5,652 feet lower in elevation but only 3¼ miles easterly from the point.

A four-wheeled drive will take one to less frequented and more difficult places, but any good car will travel easily over most of the forest roads. Thus the mountains can be enjoyed by people who can't hike or do it on horseback.

Scenery, wildlife, wild flowers, new things to see—all of these offer inducements to forest travelers.

Mountain Climbing

Everyone is thrilled at the adventures of Edmund Hillary and other climbers of Mount Everest. Someone is credited with saying, when asked why he liked to climb mountains, "Well, the mountain was there!"

Courtesy U.S. Forest Service
HEADED FOR THE HIGH COUNTRY
Backpackers on the Upper East Lostine, Wallowa-Whitman Forest, Oregon. The white papoose is the only rider in the party.

Even animals like to get on high spots overlooking the ground. An elk seeks out a knoll or a high point on a ridge on which to lie down. A group of lambs will run up onto a

Courtesy U.S. Forest Service
CLIMBING PARTY ON THE WAY TO THE TOP OF THE CHINESE WALL
Overlooking the Lewis and Clark National Forest in Montana. Here, close to the Continental Divide, you may get higher, but not much.

rock pile or stream bank, only to turn and run back down, when it would be a lot easier to run on the level. In animals it may be an inborn sense of self-protection, to be better able to see the approach of enemies. Perhaps humans have an instinctive desire to climb to high places. That may be why boys persist in climbing trees, or is it the spirit of adventure?

Maybe man climbs a mountain to see what is on the other side. Curiosity. He may do the same thing just to prove his ability, stamina, and endurance. It may be only to give him

Courtesy U.S. Forest Service
WILDERNESS WALKERS, SPONSORED BY THE MONTANA WILDERNESS
ASSOCIATION, ON THE TRAIL TO GREYWOLF LAKE
Flathead National Forest, Montana

a sense of accomplishment. Or it satisfies the competitive spirit. He can do better than someone else. For those who like to see vegetation, whether the persons are trained or amateur, the way plants and trees change with elevation offers endless chances for noting and enjoying.

The forests provide dozens of mountains tributary to every community in the forest country. Some are rocky and rugged while others slope gently. Some are above timberline, others have scattering alpine species. Some challenge experts, some are suitable for beginners.

Very few require crampons, hand spikes, and climbers ropes. Oxygen masks aren't needed. Many thousands are qualified and can enjoy the art of climbing mountains suited to their skill. And yet enough really tough mountains are available to test the skill of the very expert.

Skiing

Winter sports, especially skiing, are taking the country.

Courtesy U.S. Forest Service
THE SMALL FRY, TOO, HAVE FUN ON THE HILL
No ski boots or expensive outfits here! Just used tires for sliding

Old and young have become ski enthusiasts. Millions of dollars are expended for development of ski areas. The fa-

HOODOO SKI BOWL
Thrills galore. Willamette National Forest, Oregon

cilities are almost exclusively made by cooperating groups
such as ski associations or ski clubs. In the West nearly all
the ski areas are on national forest land occupied under permit.

Safety is one of the primary requirements of the permit.
A charge is made for ground rental, depending upon the
developments and income from the enterprise. The operating organization pays for the investment, operating cost
and interest from fees for rides, profits from meals, lodging
and rental equipment.

Courtesy R. V. Emetaz
AERIAL OF OREGON'S FAMOUS MOUNT HOOD AND TIMBERLINE
LODGE
The broken lines denote ski lifts. The track is made by the snowcat which takes skiers
and weak-kneed mountain climbers well up the mountain.

Nearly every forest in the mountain country has one developed ski area, and some have several. Twenty-three operating ski areas were reported in 1964 on the fourteen national forests of Oregon. The forests of Washington had twenty-one areas developed and in use on six forests. The number increases yearly.

The lifts include D-chair, S-chair, T-bar, and Platterpulls. Rope tows, cable tows, and snowcats are used to get the skiers up the hill. The Hyak Ski Area, forty-nine miles east of Seattle on the Wenatchee Forest, has the longest and great-

Courtesy U.S. Forest Service

ALTA (UTAH) SKI AREA
Double chair lift. Mount Superior in the background

est lift in the two states. It has a D-chair lift 5,700 feet long with 2,800 feet vertical rise. Anthony Lakes, on the Wallowa-Whitman Forest in eastern Oregon, is the highest above sea level, 7,100 feet at the bottom and 7,940 at the top of the lift.

Accommodations vary but may include day lodge, overnight lodge, warming hut, cafe, cafeteria, lunch counter, ski shop, ski school, and rentals. Similar interest in skiing has resulted in comparable ski development on most of the forests in the other Western states.

Berry Picking

Every forest affords a combination recreation-utility type of use. A variety of such resources exists. Mushroom gathering is quite common and combines outdoor exercise with a prized article of food. A person should know his mushrooms. A common, but incorrect, statement is that one can eat it if a mushroom, but not a toadstool. All toadstools are mushrooms—and the poisonous quality has nothing to do with such a classification.

I knew a man and wife who made a business of gathering the foliage of Pachistima (myrtle) and putting it in cold storage. They sold it to florists for making sprays and wreaths.

Ferns are gathered for home use and make most attractive decorations. In the rain forests of the Pacific the coast fern is so common one might be accused of bringing in the weeds. But every fall dozens gather them and ship carloads to every state.

Huckleberries are the number 1 forest crop in this group of minor products. They are frequently confused with blueberries, but do not grow under the same conditions. The huckleberry has a slightly tart flavor all its own, while the blueberry has a flat sweet taste. They belong in the same family botanically. Huckleberries are widely distributed in the mountains and often are heavy bearers of large juicy fruit. It is well to reach your favorite berry patch before the bears do. Try not to be there at the same time.

Many enjoy going out in family groups to lay in a supply of berries. They make fine pies, jams, and sauce. Huckleberries and hot biscuits! Oh, boy! Did you ever try putting a spoonful on a pancake just before you turn it over?

Sometimes they are harvested on a commercial basis. On the main divide south of Thompson Falls, Montana, I saw an operator with a special machine for sorting and fanning. A crew of pickers gathered the berries and brought them to the sorter in backpack cans. A power blower fanned out the leaves, the berries were sorted and sized on screen shakers and then put into wooden crates or flats and loaded in a covered, iced truck for hauling to the city. In the same general area a group of twenty or more Negroes were picking

Courtesy U.S. Forest Service
BEARS LIKE HUCKLEBERRIES, TOO
It's OK if the patch is large enough. If he wants the patch—don't argue

huckleberries for delivery to the Dining Car Service of the Northern Pacific Railroad. Many of them were dining car employees, their families and friends combining a mountain vacation with an income.

Photo by Photo-Art Commercial Studios, Portland, Ore.
WHO WOULD NOT QUIT PICKING HUCKLEBERRIES WITH
SCENERY LIKE THIS?
Scott Lake, Willamette National Forest, Oregon

The forests of western Oregon and Washington supply many other kinds of food and ornamental greenery. Some are: kinnikinnick, salal, huckleberry shrubbery, blackberries, raspberries, elderberries, rhododendrons, and azaleas. Medicinal plants are gathered commercially, too.

Rock Hunting

One does not need to be a trained geologist to find interest in the geological formations in the mountain country. The lava flows in the Cascades west and south of Bend, Oregon, look as though they happened just a few years ago. Actually they are young geologically.

Courtesy Jack Eng, Baker, Ore.
THE BUFFALO MINE NEAR GRANITE, OREGON
In the Umatilla National Forest. Abandoned mines interest many. This one still produces.
Ore on the dump came from an underground tunnel.

The granite in many mountains shows the effects of violent earth movements in ages past.

Farther south are found the sedimentary formations, sandstone, limestone, and shale. Here are the colored walls and erosion carvings made by wind and water.

PANNING GOLD IS NOT A LOST ART—BUT ALMOST
In the depression of the thirties hundreds took to the hills to eke out a living rather than accept welfare or charity. Ernest Dorn, of Jacksonville, Oregon, is panning gold with the same pan used by his father in 1860. Here he has found a little.

NATURE DOES PECULIAR THINGS

This granite marble evidently rolled or fell into a pocket in the granite bedrock of this stream. Rapid water caused it to roll and turn but could not budge it from the pocket. When found by Ranger Ed Mackay, the nest, or basin, was eighteen inches deep and about three feet in diameter, and just as symmetrical and smooth as the globe itself. Few human eyes have ever seen this freak of nature. Lewis and Clark went near here, but, if they saw it, they failed to mention it.

THIS MONUMENT IS AN OPEN BOOK TO GEOLOGISTS

It is on Monumental Creek in the Idaho Wilderness Area south of the Salmon River.
Geological forces require a long time to do jobs such as this. The hard rock in the
boulder at the top has protected the slim spire of softer rock below.

Quartz outcroppings often bear silent witness to the deposits of gold and other minerals. Evidences of mining for ore in places are left in the form of tunnels or shafts. The search for placer gold is pictured by the hand-piled rocks beside the old sluice boxes or huge mounds of gravel left by hydraulic "giants" or the upside down meadows where dredge boats have dug to bedrock.

Courtesy U.S. Forest Service

INDIAN POST OFFICE ON THE LOLO TRAIL, IDAHO

Mounds built by the Indians to mark a route southward from this divide to fishing grounds on the Lochsa River. Lewis and Clark, on their westward journey in 1805, climbed out of the Lochsa Canyon from a point near Jerry Johnson Bar to the Post Office. In the background, a field of bear grass.

Folks still can be seen gold prospecting in the Western national forests. In the depression of the thirties, hundreds took to the hills and eked out a living this way rather than accept relief of any kind.

Forest maps bear witness of the mining days, with such names as Ten Cent Butte, Two Color Creek, China Diggings, California Gulch, Last Chance Creek, French Gulch and many more. Why is it a "gulch" in mining country while elsewhere it is a ravine, canyon, valley, or "run?"

The "rock hound" is more common in the sixties than the panner for gold. This gets in the blood even more than gold mining. Women are as fascinated as men. All manner of rocks from thunder eggs to black agates are hauled home by the trunk full. Substantial investments are made in power equipment for sawing, shaping, and polishing. Jewelry, trinkets, ornaments, tabletops and what have you are turned out in quantity.

The hunter after arrowheads, stone knives, war clubs, spearheads and other Indian artifacts finds success in the mountain meadows where the red man held encampments or where fishing or hunting attracted them in season.

Fishing

Undoubtedly fishing is the most popular form of recreation on the national forests. Often the whole family "goes a-fishin'." Many women are more devoted fishing fans than their menfolk. There is a longer "open season" than in the case of hunting. It is an excuse to leave the hot apartments and pavements. It invites everyone when the weather is at its best, the mountain roads are open, and camping and picnicking can be enjoyed.

Fish are found in a great variety of locations. Steelhead and salmon can be taken in many of the larger streams at lower elevations. The high, cold mountain lakes yield quality trout. The meandering mountain brooks produce the most edible fish imaginable. And there are all variations of fishing waters in between. They are yours for the choosing.

Occasionally a fisherman is heard to say he doesn't care to eat fish, he just catches them for the love of it. But for most people no matter how much they enjoy the hours along the water, the strategy of hooking, or the tug and pull on

Photo by Oregon State Highway, Travel Division
CANOEING AND SAILING—SOME FISHING, TOO, IF THERE IS TIME
Cascade Mountains, Oregon

the line, the extra payoff is in the eating. After a day along a mountain stream, is there anything finer than a pan full of crisp rainbows, fried over an open fire? Don't let anyone talk you into fancy ways of cooking, with perhaps a French name. Cook them alone, in butter or bacon fat. They are pretty hard to spoil, so cooked slowly at low heat, or fast so they are crisp, they are wonderful. Not so well known is the fact that trout are marvelous cooked in deep fat. Get the kettle to boiling, take it off and drop in the trout. When the boiling stops, take them out with a spoon. Um-m-m.

Photo by Photo-Art Commercial Studios, Portland, Ore.
FAST-WATER STREAM FISHING
Every man to his own liking. There are all kinds of streams. This is a view of the Metolius River in the Deschutes National Forest of Oregon.

Photo by Photo-Art Commercial Studios, Portland, Ore.
BOAT FISHING IS POPULAR ON THE MANY MOUNTAIN LAKES,
OFTEN DEEP IN THE FOREST
This lake is in the Cascade Mountains of Oregon

Courtesy U.S. Forest Service
BULL MOOSE ON THE SHORE OF HOODOO LAKE
In the Selway-Bitterroot Wilderness Area of Idaho. At least a camera is legal, with a
year-long open season.

Hunting

Although fishermen outnumber hunters as visitors to the national forests, big game hunting is literally "big business." The Forest Supervisor of the Wallowa-Whitman National Forest estimates that in 1964, 9,700 deer and 4,300 elk were killed by hunters on that forest. These figures have been coordinated with State Game Department information and are considered reliable. That is a lot of game.

There are 175 national forests and all have some varieties of big game and the Wallowa-Whitman has a lot of competition as the home of big game animals.

One has only to check with a few meat markets tributary to any game-raising forest to learn that the wild meat during and following the hunting season seriously curtails the

Courtesy U.S. Forest Service
A PACKHORSE IS INVALUABLE FOR TAKING A BULL ELK
OR OTHER BIG GAME TO CAMP
If there is plenty of snow, a workhorse will do the job. Whitman National Forest west of Haines, Oregon.

sale of beef and pork. The butchers are buried for weeks, cutting and wrapping for lockers or deep freezers. Their cold-storage hangers are glutted with venison and elk meat.

At the opening of the hunting season, I've flown over the forest hunting areas. Camps were everywhere, as evidenced by the tents and the smoke from campfires. I've heard of the "Valley of Ten Thousand Smokes" and I've seen the "Forest of Five Thousand Campfires."

At one time elk were often killed for their teeth alone. Each elk has a pair of round-topped teeth between the grinders and the nippers. In the case of an old animal, these teeth are beautifully colored and make a popular decoration when mounted. Elk were also killed for their racks of antlers in the early years of the Forest Service. But trophy

Courtesy U.S. Forest Service
WHO SAID THERE AIN'T NO BUCKS?
Take your choice. A ridge on the Umatilla National Forest, Oregon

hunters and tooth hunters are about a thing of the past. Law enforcement supported by public opinion has corrected the practice. No doubt the price of T-bones has been a factor.

Big game hunting was the sport of kings and still is. The visitors to the forests are the kings. There is something special about hunting, and I like it best when snow is on the ground. The crisp air, the tracks in the snow, the sense of adventure, the knowledge that the story will be different every day and that anything can happen—all build up a keenness not at all like anything in town. The snow tracks can tell you of all kinds of woods dramas if you will learn. Even without the extra dividend of snow tracks, the average hunter just enjoys being out there. For the time, he has stepped back in history when a man went forth nearly every day to get his food. The cares of civilization drop away. Was that a deer? Look at the swoop of that hawk! A noise down there—better get behind a tree! What snapped the

Photo by Bureau of Land Management, U.S. Dept. of the Interior
A BUCK AND A DOE—IF YOU HAVE AN EAGLE EYE
Typical winter deer range in central Oregon south of Bend. Open yellow pine timber, bitterbrush, sagebrush, and Idaho fescue grass.

twig in that thicket? Every little thing you can learn about the vegetation, the wildlife, and the area around you adds to your day's enjoyment. It really doesn't matter much whether you get a deer or not.

CHAPTER TEN

WILD USERS OF THE FOREST

Elk

I WOULD DISLIKE TO HAVE TO DECIDE BETWEEN ROCKY Mountain elk and mule deer if only one of the two could survive in our forests. The elk is a marvelous animal any way you take him, whether to look at, to hunt, to eat, or to listen to. There just isn't any other sound that compares with the bugle of the bull elk. Possibly a lonesome Jersey bull comes closest to it if you could eliminate the bellow part and just hear the final tones. The bugle of the elk does something to one. Once I was riding up a trail, south of the Lochsa in Idaho, when three bull elk began bugling. One was a short distance ahead, one was across a small canyon and the third back down below. Their voices were much alike but different distances and positions distinguished them. They were really challenging each other. Many men would pay real money if they could have sat in my saddle that day. To me it was a worthwhile fringe benefit.

I've Indian hunted to within twenty feet of a magnificent bull during the rutting season. Nine or ten cows of his harem were close around. Ranger Arzy Kenworthy and I, on the high summer range, closed in on a mixed bunch and got so close we could almost lay hands on a couple of calves.

One day, when hunting in the North Fork of the John Day drainage, I'd tracked three elk and alarmed them in a timbered basin as I came onto a ridge overlooking the depression. I dropped back and went up the way they had gone, but on the lee side of the ridge. After going a half mile I more cautiously returned onto the ridge, spotted a young cow on a slight point, and by getting a big house rock between us, was able to come within a hundred yards of her without detection. She stood perfectly still for minutes, fully alert, then almost too quick for the eye to see, she'd turn her head sixty degrees and again remain perfectly motionless. After watching her for some time I managed to withdraw without disturbing her. Later in the day I came through the basin from above and found their empty beds. The other two had stationed themselves so that each occupied a minor outlook point fifty yards apart. Between them they had provided almost perfect detection against approaching enemies.

It's unbelievable how a bull weighing eight hundred pounds and sporting a big rack of horns can slip from sight so rap-

Photo by Bob Bailey, Enterprise, Ore.
FOUR BULL ELK IN THE RUTTING SEASON
Wallowa County, northeastern Oregon

idly and quietly. They use sight, hearing, and smell to alert themselves. To answer those who fear the elk will be eliminated I'd say, "First we'll have to have smarter hunters or dumber elk."

Jackman adds this comment: "I was hunting once on Meadow Creek in Umatilla County, Oregon. Eight inches of snow had fallen the night before—ideal tracking weather. Just after daylight I shot at a young bull and tracked him all day without catching a glimpse. I heard him many times, for it didn't take him long to find that someone was following him. That day gave me great respect for elk. To throw me off the track he used every trick that an intelligent man might use. When he came to a deer track he would step in the hole in the fresh snow made by the deer; when he came to a high log he would go forward some rods, then return stepping in his own tracks, then would jump the log; he would go into a lodgepole or fir thicket, make a perfect figure

Courtesy U.S. Forest Service
A BULL ELK TAKING HIS EASE ON A FARMER'S HAYSTACK
He preferred not to leave. This is a deceiving photograph because the elk's head shades
a part of his body. Baker Valley, Oregon.

eight and follow his own tracks around and around, finally leaving them on other tracks. By night he was tired, lying down to rest every two hundred yards, but I was lying down every hundred yards."

I was hunting alone a few miles from the Powell Ranger Station on the Lochsa River. Upon topping a slight rise overlooking a small basin, I heard an elk and glimpsed it in the brush. It was coming my way. So I dropped back a hundred yards and got behind a big four-foot Douglas fir tree. Then with rifle at the ready I waited, expecting to get a broadside shot whichever way the elk chose. Soon it could be heard breaking brush and breathing hard from its climb out of the basin. But instead of going a sensible shooting distance one side or the other of my tree, the sound indicated it was headed straight for it. Of a sudden a cow showed up right beside the tree and for a thousandth of a second we looked eyeball to eyeball at a distance of less than five feet. With a terrified snort she completely reversed her direction and in the batting of an eye completely disappeared. I didn't fire. Cows were legal kill. But what a picture. Her black muzzle was covered with beads of moisture. Her eyes were big and round and chocolate brown. Her ears and hair and whole appearance indicated thrift and vigor and alertness. That picture was worth far more than meat to me. Perhaps that statement is influenced by the fact that upon going again to the crest of the ridge I saw a spike bull elk calf in the basin and dropped him with a long second shot. That was the finest game meat I've ever eaten. It was beyond the veal stage but was unusually tender.

Jackman tells of a time when he didn't shoot. "I was hunting in Union County, Oregon. The woods were full of hunters so I went far south into some steep country without fresh man tracks. I stopped to rest just inside a dense fir thicket. I could see through open woods to the front and both sides. In about twenty minutes I heard something coming into the thicket behind me. I turned cautiously and

ELK MOVING FROM NIGHT FEEDING ON RANCHES TO SHELTER AND REST IN THE TIMBER

Bitterroot National Forest, Montana

noiselessly and soon could see some legs coming toward me. The animals wandered from side to side, exploring a little, then, apparently satisfied, an elk cow and calf lay down within fifteen feet of me. I could not be sure that they were elk until they were on the ground. They were facing back and did not see me. Somehow I couldn't shoot an animal that I could have almost touched with my gun."

But such chances are rare with elk. Usually, if you get one you earn it.

Mule Deer

This variety of deer is the chief representative of the deer family in much of the Northern Rocky Mountain area. It is also the largest. It not only weighs more but is more powerfully and ruggedly built. It is steel gray in winter.

Courtesy E. R. Jackman, Corvallis, Ore.
ELK CALF. THE SPOTS ARE NATURE'S CAMOUFLAGE
Data on the origin and location of the picture have been lost

Its antlers, shed annually, have a distinctive construction. They fork and the forks fork. There is no single main beam.

Mule deer live in timbered country but they seem to like big open sidehills or old burns. They go to the high country in summer and as the snow deepens they work to lower elevations. In early spring they tend to concentrate on the low open foothills, using the earliest grass and browse. They go higher as the snow disappears and the tender vegetation comes on.

They are a fine game animal, wary, but not as difficult to bag as elk.

One fascination of hunting is that on every hunt something unusual happens. Jackman reports that he has shot two buck deer that had no inside wound. In both cases the

Courtesy Lakeview, Oregon, Chamber of Commerce
FOUR POINT MULE DEER
Large ears and black-tipped tail. Fremont National Forest

bullet hit a rib, glanced off, and the shock had knocked the deer unconscious. It was not until the deer were skinned that he learned that they had died from having their throats cut preparatory to cleaning.

That brings up the point of guns. Guns for mule deer or larger animals should be large enough so the bullets will carry and penetrate. Small guns result in too many wounded deer that get away.

White-Tailed Deer

A white tail doe deer, I think, is the most dainty, attractive animal in the forest as she steps gracefully, one foot at a time, through the vegetation, nipping a bud here, a leaf there. She makes a beautiful picture, too, standing at the water's edge, glancing about before dropping her head to drink. Even the bucks have a long slim body and running gears. They are tan color, and when they run the bushy tail waves jauntily. The white underside gives a definite flag effect in flight. The antlers of the buck, like those of the mule deer, are shed annually but they have one large main beam, with spurs or points extending from it.

The white tails are widely distributed both in the West and in the Lake States. They are a fine game animal, more difficult to hunt because they depend on hiding and stealth. They prefer the shelter of thickets and dense undergrowth and brush. They are less hardy than mule deer and more subject to winter casualty but are prolific breeders.

An outstanding example of overpopulation was on the Kaibab Forest in Arizona north of the Grand Canyon. That was in the summer of 1919. Seven thousand head of cattle were using the forest and for every cow seen I swear I saw two deer. Some estimates placed population at twenty thousand, while others were even higher. They had multiplied to their own detriment. Hunters were few and made no dent in them. Nature, as usual, tried to handle the situation. Cougar took over. It was a cougar's heaven. There were rimrocks to hide in, limby trees to climb, and venison galore. In the evening after sundown the deer came out of the timber to feed in the openings. These parks were grazed like an overstocked sheep pasture, yet we would see 50 or 60 on a park of twenty acres. There was no browse left anywhere, while browse and flowers and weeds are the preferred feed of deer. The timber was yellow pine, which normally reproduces plentifully. Yet here I found no young tree growth at all and the needles on the larger trees were cropped as high as the animals could reach.

Moose

Moose are especially at home in the swampy lake country of northern Minnesota.

They seldom run but have a clumsy lumbering trot that enables them to move away very fast. Their large head with huge Roman nose and broad muzzle gives them a prehistoric look. Large spreading hoofs with really large dewclaws en-

Courtesy U.S. Forest Service

DINNERTIME
White-tailed doe and fawn

able them to travel easily across sphagnum swamps and mud lakes without bogging down as a horse or mule would surely do. Many times we'd go down on our hands and knees in a sphagnum swamp and get a clear drink of water from a fresh moose track.

Moose flies, large flies like horseflies, but bigger, drive the moose into shallow lakes. They like the tender water grass growing on the bottom. Occasionally a moose would be almost entirely under water except for the shoulders. Then the head and neck would emerge, the moose chewing on the grass with long stringers hanging down from both sides of the mouth and the water running down the big ears. It was a silly sight.

In one day's canoe travel my partner and I counted nineteen moose. Once we paddled between a big moose calf and the shore. He headed for the opposite bank, we after him. It was nearly six hundred yards away. When almost there, he apparently remembered his mother he'd left on the south shore. He swapped ends and started back. We did the same. Paddling as furiously as we could we were just able to keep up with him. A kodak picture at eight feet showed him breasting up eight inches of water.

Buffalo

Although not a forest animal, the buffalo on the Moiese Bison Range afford a good example of management. In the winter of 1925 I went from Polson, Montana, to Ravalli. It was on a morning following the fall of a foot of new snow. In that part of the Bison Range reserved for winter use at least one hundred buffalo could be seen from the train window. They were moving, all in one direction, grazing and pushing the snow ahead with their noses and faces. They don't paw for grass like a horse, but root ahead as they graze. Each of dozens of leaders showed up at the front end of a fresh trough—resembling a magnified pine beetle in his gallery.

Buffalo were formerly a migratory animal; whenever some individuals were pushed out of their own territory and into a place where migration was hard, such herds eventually disappeared. For example a young healthy herd once existed in Harney County, Oregon. An unusually hard winter killed all of them, years before white men arrived.

Mountain Sheep?

The animal pictured below is neither fish nor fowl. His mother was a white-faced, hornless domestic ewe sheep. She was one of a band of 1,200 head ranging in central Nevada halfway between Austin and Tonopah. She grazed on the Toiyabe Forest in summer. But there are rumor and speculation about this animal's pappy. The herder reported that a buck deer had visited the band the previous fall and spent some time in company with the ewes. From this knowledge both the herder and the sheep owner firmly be-

Photo by Charles Simpson, Baker, Ore.

WHAT IS HE?

Born of a white-faced domestic sheep, it is not known who his father was. He is probably half bighorn mountain sheep.

lieved that the unusual character was half sheep and half deer.

His horns could pass as those of a yearling Rambouillet ram but there were no other similarities whatever. They do resemble very much the horns of a wild mountain sheep of the same age. They have no similarity with a deer's antlers, that are shed and replaced annually while this ram did not shed his horns.

His hair—you couldn't call it wool—would readily pass as mountain sheep hair and was far different from either deer hair or sheep's wool. His bare belly and his seven- or eight-inch tail relate him neither to a deer nor to a sheep. A brownish-tan patch over his hips gives a hint of mountain sheep color. When he was loose among the band of sheep he stood out as "different," especially as he was at least eight inches taller than his domestic associates.

While making no claim of being a biologist, I have to conclude the chap was half mountain sheep and half domestic sheep. The two animals are not so different that crossing would be impossible or improbable. The fact that such crosses appear to be extremely rare could be explained by the fact that there are so few mountain sheep in the mountains, and domestic sheep are normally out of the summer range before breeding season.

The herder told me five of these "different" lambs were born but only the one lived more than a few days. This animal had not had an opportunity to prove his capability as a father. This is a true story. I felt of his hair, and I looked him over.

Cougar

While on the Kaibab Forest in Arizona, an overnight stop was made at Uncle Jim Owen's cabin. Uncle Jim was a cougar hunter. And old "Lead" was his cougar hound. He claimed a record of 136 cougars killed. His story had been told in the *American Magazine* some years earlier. Teddy Roosevelt had come to the Kaibab and hunted cougar with him and had given Uncle Jim a pearl-handled pistol with "T. R." inlaid in gold. He prized that gun beyond money and some scoundrel had stolen it not long before my visit.

In the winter of 1928 I took part in a cougar hunt on Fish Creek, a tributary of the Missoula River in Montana. A merchant in Alberton was the ringleader and a fireman on the Wallace branch of the Northern Pacific was another promoter. Ben Vogel, a trapper-hunter for the Biological Survey was the expert and the owner-handler of a trained cougar dog. A chap from Missoula and I completed the safari. Plans were made in advance. One morning following a new snow of eight or ten inches, word flew around, "To-day's the day."

We gathered at Alberton. We backpacked our beds, groceries, guns, and equipment about eight miles up to the Fish Creek cabin. This was a wintering ground for white-tailed deer and we encountered scores of them en route to the cabin. We still had time to cut a supply of wood, cook up a big feed, make our beds and get our outfit ready for the big hunt the next day. And the whole conversation was about cougars, deer hunting, dogs, game, and treeing. One of the guys proposed that if we got a medium small cougar we tie him up and take him back to town alive. This raised the question of how to get him roped in the tree and down to the ground. Next problem was getting his feet tied and a muzzle on him. Then he had to be backpacked all the way out and all the packsacks were too small. Should we make a sled and drag him?

We had some ropes in our outfit and others were hanging in the cabin. But were they strong enough or long enough? Then some leather straps were located. Two of the party were down on their knees on the cabin floor tying knots and cutting and fitting straps, striving to make muzzles and halters and hobbles. Some of the rigs were too big, and others were too small and had to be worked over. Of course everyone had in mind a different-sized cougar. By midnight Ben had boiled up a big feed of whole wheat and fed

Spot his day's ration, and the rope splicers figured they were all set for the capture.

After a few hours sleep we were off before daylight with the rigging, guns, dog, and lunches. Deer were plentiful and tracks in the twenty-four-hour snow looked almost like a band of sheep had been there. About nine o'clock we found a cougar track and Ben turned the dog loose. Nose to the ground he was off with the hunters behind as best they could. He tracked until a buck jumped in front of Spot who, for reasons known only to himself, quit the cougar track and followed the buck. Ben finally caught him and gave him a trouncing. He claimed the dog had never chased deer before and he had to break him of it. We suspected he was so mad and embarrassed because Spot had let him down that he was sorely tempted to shoot him.

Fox

The sight of a fox in the wild is a real occasion. Near Mount Diablo on the Elk Summit District, Idaho, I was walking uphill in open brush land and had stopped to get my wind. Suddenly a red fox walked noiselessly into view and stopped in an opening. He was about thirty feet from me, but didn't see me. He was hunting and intent on some rodent or bird. I remained motionless and shortly he went on, crosswise to my line of travel without being alerted. That was one of the most striking pictures I ever saw. His inimitable red fur was shiny and in perfect condition. His ears were erect, his eyes were jewel-bright and he carried his graceful plume regally.

Wildcats

Bobcats are a frequent sight in the foothill ranges. They look much like an oversized house cat with a stub tail. They do a little damage by killing birds or domestic chickens, or even lambs at times. The score is balanced by the disposal

of mice, squirrels, and other culprits. I've never molested
them and think they deserve protection.

The big gray Canada lynx is comparatively rare. I over-
took one on a narrow mountain road in the Coeur d'Alenes.
He started off ahead of me, my car lights providing a ring-
side view of him as he loped along easily. He stood tall and
laid those big furry feet down in easy bounds. I pressed a
little harder on the gas pedal. The speedometer passed thirty
miles, which was about as fast as was safe on that sidehill
fire road after dark. He kept in the lead for more than a
quarter of a mile, then decided he'd do better in the rough.
With a super diagonal bound he made the top of the five-
foot back slope and was lost to sight in the dark.

Porcupines

Porcupines were protected for years by an unwritten law
because they were considered a lifesaver to a person lost in
the woods. Actually, saving by most any other way would
be preferable. Besides, they are nature's worst pest. The old-
fashioned pork barrel was one of their favorite dishes. Any
outdoor toilet was a treat, too. Jack says he once came across
a mountain cabin with a detachable seat that hung on a
nail in the outdoor toilet. Under the seat was written:

> Blest be he by Heavens crew,
> Who hangs this seat when he is through.
> But cursed be he to Hells dark maw,
> Who leaves it down for things to chaw.

I've had axe handles, canoe paddles, canoe seats and other
items ruined. They seem to crave brine, sweat, or any form
of salt. They raided our camp kitchen one night and the
commotion made one think a bear was in camp. They can
be the death of cattle, horses, and dogs. It's pure hokum
that they can throw their quills. They slap their tails at
molesters and leave a load of quills. I once had a colt tangle

with one. We had to throw and hog-tie him. With pliers we pulled them from lips, chin, nostrils, and both front legs from hoofs to knees. The quills we pulled were an inch deep in the bottom of a water bucket. To make matters worse the porkies peel and girdle young trees. Damaged trees either die or become forked or spike-topped. Foresters have now declared open warfare on the pests. Strychnine and brine-soaked boards seem to work best. A pistol will scarcely do the job. They're harder to kill than a lion.

Pack Rats

Pack rats can also be classed as troublemakers. They take possession of any and every abandoned cabin. Their imagination is good, too, because they often take over before the owner has decided to abandon the building. Once in possession, the odor, the droppings, the sticks and chips, the grass and leaves are sufficient cause for any but the most callous owner to let them have it. One time I spread my bed in a cabin at the Yellowjacket mine in the Salmon River country. I was awakened by a sharp pain in the lobe of one ear. I threw up my arm and knocked away a big pack rat. Blood was oozing from the ear that was swollen and tender for days. That same night a pair of socks, with garters attached, disappeared. The rascals evidently had a liking for the metal of the garters and took the socks along for good measure. They prize anything that shines such as pieces of tin, glass, or metal, and their nests are often loaded with such trinkets.

The favorite method of control is to arise about midnight, walk barefooted on whatever is on the floor or ground, button up your underwear, hold flashlight in left hand and pistol in right. The only better way is to let your partner take the light and do the walking while you stay put on a warm spot and train a .22 rifle on rafter or joist.

You can't trap them. At least one expert tried it. A two-foot joint of stovepipe was nailed on a table with one end projecting out into space. A trap was set in the pipe and an apple placed at the outer end. Others were placed as bait along the table. In the morning every apple was gone, including the one in the stovepipe but beyond the trap. The trap was still set.

On a pack trip north of the Grand Canyon of the Colorado, Jack Roak and I had a rat experience. We had some rice that came in a cereal box like cream of wheat. We had used rice for one meal from the container. Upon pouring from the box for a second meal, whole oats flowed into the kettle instead of rice. Every kernel of rice was gone and the box was full of oats. We had a sack of oats for the horses and it was twenty feet from the grub box. The hole in the box was about the size of a man's thumb. We found no trace of the rice and never saw the trade rat. Evidently he liked our rice but didn't want to make us mad, so he laboriously moved enough oats from the grain sack, kernel by kernel, to fill the cereal box. His conscientious nature was not appreciated.

On the same trip we saw dozens of white-tailed Kaibab squirrels. They occur only in one other place in the world. They are a size larger and the same color as pine squirrels. But their plumy tails are pure white. When they jump or run, or even sit still on a limb, the white plume gives them away. The marvel is that this has not led to liquidation by enemies long ago.

Rattlesnakes

Rattlesnakes should be on the decrease because nobody loves them and everyone kills all he sees. A person seldom finds rattlers in high mountain timbered country, but every forester has seen and killed his share. I have killed scores but only one attacked me. I was walking down a hill through an aspen grove when I heard him rattle behind me, slightly to the right. He was coming at me on the double, his front third up in the air. His mouth was wide open and his forked tongue was going like a sewing machine. A planting mattock quickly dispatched him. It wasn't clear whether I'd walked over him or real close to him but he was really mad. Nor do I know why he hadn't struck me.

On another occasion I'd killed a great big rattler. After he was stone dead my research instinct got the better of me. I'd heard about venom sacs and I wanted to see one and see what the fangs looked like. With my jackknife in one hand and pencil-sized stick in the other I opened his mouth and did some probing within. The panic button or something must have been pressed. Two nice streams of venom arched up and out about six inches and landed squarely on the back of one hand. I had no idea I could move so fast. I had no breaks in the skin right there, so no harm resulted. Scientific interest waned in this researcher.

Game Birds

Blue grouse is the number one game bird of the forests. They are large birds, able to fend for themselves quite well, and are good reproducers. They have large breasts and their meat is delicious. Much of their time is spent high from the ground on sizable limbs of trees. Their blue, gray, and brown coloring blends well with the bark of fir, and they remain motionless. Flocks of eight to a dozen were often seen in a single tree. If a hunter with a .22 rifle picks off the bird lowest down in the tree, the others will seldom fly. I have even used a .30-30, shooting at the heads, and have knocked down three or four birds before the ones above took flight.

Fool hens are interesting but it would be a shame to kill one. They have been scarce for years. They, too, will sit perfectly still but on low limbs three to eight feet from the ground. They could be knocked down with a club or even grabbed by hand. Predators would probably keep numbers limited even without man's help.

Ruffed grouse, also popularly known as willow grouse, or fantails, were a popular and numerous game bird fifty years ago. They have about half the body weight of blue grouse but are excellent eating. They run about on the ground, among the willow clumps or other brush. They are adept at getting out of sight. Now you see them and now you don't. And if they fly it's only a few rods to another thicket. A .22 rifle is the weapon to use but you've got to be quick on the trigger.

In recent years the imported Chinese pheasant, Hungarian pheasant, and chukar have come out front as game birds. They are numerous and more accessible and have largely taken the place of the native grouse as game birds.

Rabbits

On Thanksgiving, 1913, Nils Eckbo and I were at Frazer Ranger Station east of Rogerson, Idaho, and decided to go hunting. Nils had a big French horse pistol that shot a single

Courtesy U.S. Forest Service
A LIMIT OF BLUE GROUSE
The number one game bird of the western forests

ball, about the size of an ordinary marble. Soon we got on the trail of a big mountain hare. We followed him up a draw some distance, then he climbed out on the sidehill, swung back behind us, then down the draw. It was snowing steadily so we could track him even when he traveled in his earlier trail. He made a loop from down the draw and back into the old trail above us, thus making a figure eight. After making the second double loop, we managed with our combined scheming to outwit him. I took a position at the forks of the figure eight while Nils followed the tracks around the upper loop. On the back track Mr. Hare stopped when he spotted me and gave Nils a chance to drop him with one ball when he came within range. Perhaps some animals don't use their brains for thinking purposes, but sometimes you wonder. It is known as instinct with animals, intuition with humans. But I have seen animals do things that seem to go far beyond mere instinct. Both words are defined as action without thinking.

Fur Bearers

The present population of fur animals is only a fraction of that of fifty years ago. They included beaver, otter, mink, muskrat, fisher, marten, weasel, and skunk. Remnants can still be seen by persons spending much time in the forest. Pine marten are about the most numerous with beaver a poor second.

Professional trappers are nearly a thing of the past, too. One chap, a summer Forest Service employee, decided to become one. He stocked up on groceries, traps, and winter clothes and with a young bride packed far down the Lochsa River to a one-room cabin he had put together. He next laid up five or six lean-to shelters for overnight stops on his trapline. There were no neighbors and no human visitors. Both contracted a severe case of "cabin fever" and, after a too-long winter, they came out to the settlement neither speaking to the other nor giving any sign of knowing the other was present. He went one way and she the opposite. Monotony is the rock on which many marriages have foundered.

These animals of forest and stream have distinct historical interest. When one is discovered by an observant forest traveler he gets a real thrill.

Bears

What is a predator to one person may be a real game animal to another, and to a third a wonderfully interesting fellow creature. So much depends upon the interests of the person doing the classifying. The black bear is a prime example. Not only do individuals fail to agree, but the legislatures and game commissions of different states have differing opinions. The status of the black bear sometimes changes in a single state. He may be a game animal one year and a predator the next. A brown bear is also a black bear. The naturalists say so and most of us agree, having seen a black bear nursing a brown cub and a black and a brown cub being disciplined as brother and sister by a brown she-bear. Oblivious to classification quarrels, he does pretty well for himself, is widely distributed, fairly numerous in most forested areas and yet never seems to overpopulate the range.

We do know that bears raise havoc with bands of sheep. It is generally the supposition that this is due to killers that start on a sheep diet and get the habit. Of course any bear having a joint territory with sheep may eat a dead sheep or a lone crippled sheep and thus get the taste for mutton. When a bear becomes a killer there is no other remedy but to trap or shoot him. As a preventative it is quite common to keep a rifle handy and use it on any bears found on the sheep range.

Bears have a wide variety of eating habits. You are likely to find one or two in any good huckleberry patch. You've seen large rotten logs torn to shreds where bruin was getting a feed of woodworms, ants or ant eggs. They are fond of fish and some become skilled fishermen. They prey on small rodents and eat carrion that results from many causes. They examine garbage pits with thoroughness and may molest camps. But except the sheep owner, who can say that the bear does more harm than good?

Most bears killed by sportsmen are those encountered while the hunter is after other animals. Some are taken for their hides as rugs or wall trophies. The meat is edible but is fat, even more so than pork, which it resembles. Bear meat is a bit repulsive to many for two reasons. First, bears are known to be promiscuous in their own choice of food. Second, a bear carcass, skinned and drawn and hung up on a gambrel stick looks very much like that of a human. The legs hang down loosely and do not assume the rigid pattern of a dead steer.

The black-brown bears are usually harmless and will vanish in the woods upon the approach of man. This does not mean they will not fight if cornered or will not harm persons teasing or feeding semi-tame specimens in national parks. And mother bears with small cubs are dangerous and should be avoided. One time I was riding along a timbered trail when I encountered a big brown bear. With a furious growl she appeared ten feet from the trail. I had difficulty preventing my horse from leaving those parts. The bear went right along with the horse about three paces to the rear and the same distance from the trail. She kept up a continuous snarl and chomped her teeth until she had white foam all around her mouth and face. All of us were concerned. The horse wanted to leave the bear behind right now. I was concerned with not being left in the dust of the trail with the mad bear. She wanted to push the horse and me out of sight and sound of her cubs. After five hundred yards she decided she had made her position clear and terminated the pleasure of our company.

Jim Hutchens, Fire Control officer on the Glacier View District of the Flathead Forest in northwest Montana, provides the following:

"Sometimes bears give us trouble in our camps and cabins. At one cabin we kept our ham and bacon in a wooden box that had been covered with tin. This was fastened to the end of the log cabin with heavy wire. One morning after we had all been gone for the night we found that a bear had gotten the box down and open and had feasted on our ham and bacon. We put it up and refilled it and that night about 1:00 A.M. the cabin began to shake and rattle.

"We ran out and there was the meat box on the ground. I had a pistol and fired a shot in the direction of the bear who was standing about one hundred feet away in the moonlight watching us. He ran off and we put the box back on the wall and went back to bed. We thought we could go to sleep and not worry about anything the rest of the night. But he was at it again twenty minutes later.

"We took the box, meat and all, into the cabin and were not bothered any more that night. A few nights later, we had to eliminate this bear as he would not leave us alone."

Coyotes

Coyotes are regularly classed as predators. They are widely distributed and are found in the high mountains, in forested areas, in the foothills, on winter ranges, and in farming areas. They are a serious menace to the sheepmen, sometimes killing scores of lambs in a band of sheep. Sometimes they even kill newborn calves right around the corrals of isolated ranches. They are death to young fawns and will even gang up on a grown dear and kill it in late winter in deep snow. They take the eggs and young of some game birds and lie in wait for poultry.

It is true they are good hunters and kill many mice, squirrels, gophers, and other objectionable critters. Still, the count is not in their favor, and the relentless warfare against them continues. They have been taken for their pelts. Bounties have been paid for their scalps. They have been condemned to death on sight by farmers, sheepherders, and hunters. Government employees of the Biological Survey, later the Fish and Wildlife Service, were hired to eradicate them. They have been trapped, shot, strychnined, and gassed with cyanide guns. A peculiarly effective poison known as "10-80" was developed and used successfully. The latest is a means of preventing reproduction. But so far the coyote persists and no doubt some of his folks will be with us quite a while.

Birds

Eagles, hawks, and a few other birds are often considered harmful. Naturalists have been able to show that certain varieties of hawks are predators while others are definitely beneficial. To the vast majority a hawk is a hawk and a good species is just as likely to be shot as a bad one. I've seen a hawk kill and carry away a Chinese pheasant hen probably outweighing the hawk.

Eagles are known to kill and carry off lambs. An eagle will occasionally kill a fawn.

Crows are thought of as pests, especially in a newly planted cornfield. But recently I counted eleven crows in a pasture that was being irrigated. Pocket gophers were numerous and being forced above ground by water. A crow would attack, striking blow after blow with his beak until the gopher gave up. Then the fur would be stripped and flesh eaten a chunk at a time, or the whole gopher carried away, no doubt to a brood of young. So perhaps nature does a pretty satisfactory job of keeping things on an even keel, and man should take a hand only in clean-cut, specific cases of known killers. The difficulty is that the man fighting for the predators is not the man who suffered the losses. This goes way back to biblical days.

Game Management

As our Western population increased along with hunter

participation, game problems developed. The relatively new
profession of Game Biologist came into existence. The states

FRANKLIN GROUSE OR FOOL HEN
This is a mature male

became first licensing agencies, then protective organizations, and last game managers. The game management business is complicated and difficult because of a lack of control of many of the factors involved, and because of conflicts of interest.

By comparison, managing a large cattle ranch is simple. The owner can adjust the number of stock to fit his forage supply. By one of several means he can balance his summer capacity with his winter feeding needs. He can confine his stock to winter quarters until the spring range is ready for grazing. He can absolutely control the sex ratio, cashing in on his steer crop and using a maximum amount of his forage for the female breeding herd. One big advantage is in being able to market what and when he wishes. He may dispose of the fat dry cow, the old-timer, the barren heifer, the poor milker or mother. He doesn't have to convince anyone that he's doing the right thing, unless it may be his wife or his banker.

But the biologist first has to look hard at the conditions. This calls for weeks and months of riding and hiking, summer, fall, winter, and spring. "Conditions" cover a multitude of things. What are the trends? Can more game be produced? Are there too many now? Are the males too numerous or too few? What about winter loss, present or potential. What about food supply for each season? Then the question arises, what needs to be done? The game animals will likely go on doing about as they have been and it's not always easy to induce sportsmen to agree to needed changes.

In the Coeur d'Alenes, a mining district with Wallace, Mullan, and Kellogg as centers, a high percentage of the male population worked underground and they were mainly active hunters and fishermen. The deer population was low considering the numbers of licensed hunters. But deer of either sex could be taken in the thirties. This meant a continual drain on the breeding stock. But a person would almost be helped out of town on a rail at the mention of killing bucks only and they made it stick. On the other hand these same folks had one of the most active sportsmen's associations I've known. The Shoshone County Sportsmen's Association, for example, arranged for two truckloads of elk (about sixty head) to be brought from north of Yellowstone Park to their territory. The Forest Service, using CCC crews, built two big, high corrals and fed hay from midwinter until spring when the elk were turned out. The association provided the hay and paid the entire cost of delivering the elk. All the elk were ear-tagged. That fall one of the tagged animals, a young cow, was shot on the Clearwater River, having crossed three rivers and intervening main ranges of mountains.

At that same period the same sportsmen's association built a fish hatchery and rearing ponds and turned loose in the local streams legal-sized trout. They were in advance of their own state game department on this score. Ellis Hale, the president-manager of the Coeur d'Alene Hardware Company, of Wallace, was a leading sportsman.

In some places the sentiment about killing does is just the

YOUNG ELK TRANSPLANTED FROM NEAR YELLOWSTONE PARK TO SHOSHONE COUNTY, IDAHO, BY THE LOCAL SPORTSMEN'S ASSOCIATION
Coeur d'Alene National Forest, February, 1939. CCC enrollees built the corrals and fed the hay provided by the association.

reverse. Even where the deer population is large in comparison with the available winter ranges, or where the game is doing serious damage to ranchers' and farmers' property, the sportsmen are opposed to the slaughter of females. One of the arguments is that if hunters do not have to look for horns, more accidents to hunters will occur. Others seem to think that more and more deer are needed, so why chop off the brood stock. Others are opposed to killing the does for sentimental reasons. Sentimentalists are always cropping up advocating non-killing of game. The cruelest death is the long-drawn-out torture of starvation. Many states have found the female to be the best management tool of the game biologist. Save the does or cows to build up the herd —utilize the does or cows to control or reduce the herd. Application must be herd by herd or unit by unit.

As mentioned under the part dealing with white-tailed deer, a serious situation existed on the Kaibab Forest in 1919, in that section of Arizona lying across the Grand Canyon from most of the state, with few resident hunters. The nearby Utahans had plenty of deer in their own state. The climate was favorable for game. Cougar were the only enemy and were far from a control factor. What few sheep had grazed on the forest had been removed in earlier years. It appeared that the cattle would have to go too. The state authorities did not have a solution. To save the vegetation and the deer herd itself the Forest Service finally took steps that I'm sure had not been done previously or found necessary elsewhere. Permits were issued to shoot deer and remove the meat, without game licenses and regardless of residence. Publicity was given. Extra forest officers were sent to the area to supervise the hunt.

Pennsylvania is a wonderful example of what game management can do. It once was a hunter's paradise, but closeness to huge population centers caused a serious shortage of deer. Complete protection of does reversed the situation and at last reports the state has one of the biggest deer herds of any state. On a recent trip through the state, I saw frequent signs on the highway saying "Caution—Deer Crossing."

CHAPTER ELEVEN

WHITE PINE BLISTER RUST

WEBSTER SAYS A FUNGUS IS A "CRYPTOGAMOUS PLANT, DES-
titute of chlorophyll and deriving nourishment wholly or
chiefly from organic compounds." Blister rust is one of those
things. It produces no true flower but propagates by spores.
As with wheat rust, but unlike most other plants, it lives on
two entirely different hosts and has two dissimilar spores.
Spores are produced by many fungi, and by a few plants
such as ferns. They are not seeds because spores are frequently one-celled while seeds are amazingly complicated
affairs. But spores serve the same purpose as seeds.

One form of the rust spore is produced on trees and is
carried long distances by the wind. If this spore lands on a
leaf of a currant or gooseberry bush it grows and forms a
leaf spot or leaf rust. Both currants and gooseberries belong
to the genus Ribes (Rye'-bees) and that term is often used
by those working on blister rust.

The spores produced on leaves of the Ribes mature and
they, too, are carried by the wind. But they are more delicate and will survive only a short time. Their goal is the twig
of a tree. If they get there soon enough they attach themselves to the tender bark, grow, and produce spores that repeat the process.

These spores from the Ribes bushes are not only delicate
but selective. They don't want just any old tree. They seek
only pine trees and not any old pine tree. Peculiarly they
will grow only on a pine with five needles to the cluster.

They will pass up a ponderosa pine completely with its three needles and ignore a two-needle lodgepole like a tramp on the road. Any five-needle pine is their meat. In the northwestern United States western white pine, whitebarked pine, and limber pine are at home. All are five-needle pines. Another is the sugar pine of California and the southern fringe of Oregon. Western white pine is the only commercially valuable one of the three susceptible pines native to northern Idaho and western Montana.

The rust spore causes a swelling on the young pine twig. It produces a cancerous growth that soon kills the twig from the affected spot out, thus biting the hand that feeds it. In the meantime the fungus matures and develops an orange-colored spore-producing area on the twig. A single tree may have dozens or hundreds of these reddish or dead-brown "flags" showing in the crown. Naturally the younger

Courtesy Div. Blister Rust Control, Bur. Plant Industry, U.S.D.A.
WILD CURRANTS AND GOOSEBERRIES, GROWING IN THE OPEN,
ARE DENSE AND BUSHY, WITH AN EXTENSIVE ROOT SYSTEM
They are the alternate hosts for a fungus destructive to five-needle pines

Courtesy Div. Blister Rust Control, Bur. Plant Industry, U.S.D.A.
"RIBES" ("RYE-BEES") GROW IN THE SHADE OF TYPICAL
WHITE PINE STANDS
They are spindling and have only a small root that is readily pulled by hand. The
string line marks the boundary between the worked and unworked area. On the
worked area the berry bushes have been destroyed.

and smaller the tree, the more disastrous is the rust. The smaller trees die outright. Pole-sized trees become spike-topped, forked, and deformed. They are slow-growing and unthrifty. Mature pines will live on for years although growth is slowed up. They are excellent incubators and cause heavier concentrations of rust on the Ribes. It is a vicious cycle.

Blister rust of white pine is an immigrant. It came from Germany on young seedlings shipped to British Columbia, Canada, and got a foothold there. It then spread east and south into Idaho. First it was discovered in two small spots on the Pend Oreille Forest, and later the same year it was found on the Coeur d'Alene Forest. Spread was rapid and alarming.

In 1931 an important all-day conference was held in Spokane, Washington. S. B. Detwiler, plant pathologist of the Bureau of Plant Industry, Department of Agriculture and Regional Forester Evan Kelly, Missoula, Montana, were the principals. Five supervisors of the white pine forests were concerned listeners.

Mr. Detwiler outlined the history and nature of the rust, the seriousness of the problem, the alternate ways in which it could be combatted, and the tremendous size of the job. Regional Forester Kelly seemed to question an eradication program, but he really was insisting on being shown. He asked pointed questions; he challenged some of Detwiler's statements. Detwiler was a scientist, sure of his ground, and he pressed for action. Kelly, still skeptical, probed and dug for facts. Finally a decision was reached to undertake the work. The Bureau of Plant Industry would do the technical work and train forest leaders, and the Forest Service would do the actual eradication.

The remedy is removal of the currant and gooseberry bushes, thereby breaking the rust's life cycle. The one helpful fact is that the spores from the Ribes die quickly. The pathologists told us that they would not survive a flight of more than one fourth of a mile. If you had a section of val-

uable white pine land and eradicated all the Ribes bushes on
the square mile and on a quarter-mile strip surrounding it
you would protect the tract from rust. It would mean, also,
that if you covered 15 percent or 50 percent or 90 percent
of a forest but did not finish the job, all would not be lost.
The worked areas would be saved.

The Forest Service went all out. White pine was then, and
still is, the queen of the timber trees. Appropriations for
control were obtained. Multitudes of WPA workers were
used. I sat down on a log with one of these fellows taking
a breather. Our conversation was brief and not about Ribes.
He said he couldn't see how the country could ever get back
on its feet. I did my best to give him a lift and I hope he
survived until better times. In 1933 the CCC program
was started and the Coeur d'Alene drew eight two-hundred-
man camps the first season. This was stepped up to fourteen
camps the second year and sixteen camps the third year.

Courtesy Div. Blister Rust Control, Bur. Plant Industry, U.S.D.A.
BLISTER RUST IS ESPECIALLY DAMAGING TO THE YOUNG WHITE
PINE STANDS, WHETHER NATURAL OR PLANTED

Nearly 90 percent of the enrollees were used on blister rust work.

Currant and gooseberry bushes occur in most all locations where white pine thrives. In the shade of timber they are spindling, with no strong root system, and can be pulled by hand. A crew of about twelve men work abreast, six to ten feet apart, advancing through the timber. Each carries a specially built short-handled grub hoe for use on tough plants. The end man as he moves along tosses ahead a string ball that unwinds and serves as a marker or guide for the end worker on the return trip. This avoids the likelihood of going over the worked ground, and, more important, prevents leaving untreated strips.

The crew foreman came right behind the crewmen. He watched the performance of each man, checking to see if he knew the Ribes, was missing any, and was getting all the roots. He saw to it that no gaps were left between workers.

Courtesy Div. Blister Rust Control, Bur. Plant Industry, U.S.D.A.

HAND PULLING THE RIBES

Note the string line. The end man on a crew of ten or twelve tosses along a ball of string as he works to mark the boundary of the worked area.

The men had to be taught to identify the different bushes. There were plenty of Ribes to handle without pulling other browse plants, such as wild rose and snowberry, plentiful in North Idaho.

Qualified trained checkers made systematic spot checks on the areas after the foreman reported a unit completed. The allowable "misses" were very small and if an area did not pass it had to be reworked. Checkers had to be experts. They worked alone and their reported findings, if in error, could be disastrous or expensive.

In openings, stringer meadows, or burns the bushes took advantage of the sunlight and grew big and husky with a vigorous root system. Here the hand pullers were ineffective. Small patches were dug out by special crews using long-handled mattocks. If larger tracts of open-grown bushes were found a bulldozer with teeth instead of a solid blade was used to uproot and pile the debris with as little earth

Courtesy Div. Blister Rust Control, Bur. Plant Industry, U.S.D.A.
HAND SPRAYING OF SODIUM CHLORATE FROM A BACKPACK CAN TO KILL CURRANTS AND GOOSEBERRIES
Other vegetation can be saved. The chemical is a hazardous fire substance.

as possible. In the fall the bulldozer piles were burned, as were those resulting from the mattock work, to be sure that the bushes did not take root again. In hand pulling the roots were knocked free of dirt and, if possible, laid on stumps, logs, rocks, or off the ground so the roots would dry out quickly and die.

Experiments were started to find a chemical treatment that would kill Ribes without killing desirable vegetation. Sodium chlorate and calcium chloride, applied by means of hand sprayers with backpack cans were used successfully. With hand sprayers the damage to other plants was negligible. Extreme care had to be used to avoid fire. Spraymen got the liquid on clothing from the spray itself and from back cans and handling, and some casualties resulted. Sodium chlorate supplies its own oxygen, so a fire resulting from it can't be smothered.

In Wallowa County, Oregon, a man working with it was thoroughly briefed, but he became careless and his clothes and shoes were saturated. As he crossed the street in the town of Enterprise, a nail in his shoe created a spark and instantly he was bathed in flames. He did not run or panic but grabbed off all his clothes. N. C. Donaldson, then county agricultural extension agent, stood talking with him next day when a pretty girl passed and smiled. Donaldson asked who she was and the man said, "I dunno. Jest someone who was at the fire, I guess."

Blister rust work was tedious for the CCC enrollees, but it was healthful, outdoor employment, and they became familiar with the timber country. They learned the virtues of observation, teamwork, and thoroughness. Nearly all left stronger in body and in spirit.

The program was a success. Ribes plants and rust on the two hosts can still be found, but white pine can be counted on as a crop tree and managed accordingly. Plantations of white pine are started with assurance of survival. The Coeur d'Alene reports treating five thousand acres annually for blister-rust control. This is in the nature of maintenance work to give security to stands of young growth.

CHAPTER TWELVE

CCC — CIVILIAN CONSERVATION CORPS

THE CCC PROGRAM WAS INITIATED IN 1933, ON A NATION-wide basis, to train unemployed youth and carry out needed conservation work. Each CCC camp was set up to house two hundred men and was administered under a dual organization, army and technical agency. The army had the responsibility of feeding and housing the men, paying them and providing medical care and the educational program. That organization set up the camp, tent or lumber, and maintained the quarters. The army personnel at each camp consisted of a

> Commander
> Assistant Commander
> Medical Officer
> Civilian Educational Advisor.

The technical agency might be any one of numerous state and Federal organizations and was responsible for the work projects and the safety and supervision of the men while out of camp on project work. Their personnel included:

> Project Superintendent (called the camp superintendent)
> Six or seven crew foremen
> Specialists as needed—surveyors, checkers, etc.

The Coeur d'Alene National Forest found itself neck deep in CCC business. A few companies were assigned yearlong,

the balance moving south for the winter months. The enrollees came from widely separated areas. Two companies
were 100 percent colored, one from Harlem, the other from
Jersey City, N.J. Our biggest number came from Kentucky
and West Virginia. Arkansas and Oklahoma furnished us
several companies. Each camp was supposed to have ten L.
E. M.'s (Local Experienced Men) to act as leaders in the
crews but they were not always available or did not always
stay out their enrollment. Enrollees were paid thirty dollars a month and board and clothing. A portion of their pay
was usually assigned to their home folks.

The first year all the army officers were from the regular
army. Several were majors and most of them looked upon
their assignments as undesirable. Living out in the forest,
miles from town over dusty or muddy roads, mostly in tents
with lumber floors and frames, eating food prepared by enrollee cooks, often inexperienced, was quite a change from
well-ordered normal life at a military post. Further, they
could not depend upon army-type discipline because the
enrollees were not enlisted men, and many took to discipline
as a hen does to water. The second officer at one of our camps
was a navy man and we got a kick out of his navy language
and expressions. For example his camp got its water from
an impoundment ten feet square above camp on a little creek.
He'd say, "I've got to go up on the hill and take some sound-
ings." We suspected that a few of the officers had been given
assignments by their superiors as an easy way to be free of
them for a while.

The second year, reserve officers filled all the army's CCC
jobs. We were in the depths of the big depression and they
were glad to get steady jobs. Most of them had had businesses of their own or had handled men and could handle
camp business in fine shape. They found ways to manage
the enrollees and discipline them without army regulations.
The reserve officers really saved the day for the CCC program. Most of them were captains or lieutenants.

The usual arrangement was to have an officers' mess where

both army and technical agency overhead ate. At the evening meal at one camp, the enrollee waiting on table brought a fine-looking chocolate-frosted cake and set it down near the captain. The latter said, "Do the boys have cake, too?" The enrollee answered, "No, sir." The officer said, "Take it back to the kitchen. The officers eat the same as the boys do."

Some of the camps had quite a struggle finding or training cooks. One of the colored companies was rumored to have twelve dining-car cooks among the enrollees, but they tried out every one of them. The rumor was unfounded. One of the frequent dishes of that camp was a mixture of rice and brown beans boiled together. Either makes a good dish by itself. The combination was a bit hard to take. At another camp, as I talked with Project Superintendent

CCC CAMP F-29 ON THE COEUR D'ALENE NATIONAL FOREST IN IDAHO
This was the home of two hundred enrollees while they were in the Civilian Conservation Corps.

Lang, he remarked, "Did you ever know that the Wallace Meat Market doesn't sell anything but boiling beef?" After a moment he continued, "But they'd boil watermelons out here if they got them."

Each camp received its ration allowance at the beginning of each month and could buy what and where it pleased. Some camps, with poor buying or poor supervision of the mess, might live high the first three weeks and have to tighten their belts the last week.

The first year, one of the majors had a private tent set up for his own use and he made an orderly out of an enrollee. The orderly brought all of the major's meals to him and he dined in solitary importance.

A small stream was named Nowhere Creek and one of the camps was located on a nice flat at the mouth of the stream. Naturally it was named Nowhere Camp and this was the basis of many jokes pulled on the men from that camp.

Inspectors from Fort George Wright, Spokane, visited the camps periodically and on one occasion were looking over one of the camps on the Magee District. A big screened latrine was on the sidehill a little distance from the barracks.

It was visited first. Rolls of toilet paper had been thrown around until the place was well festooned, many of the lids were off, and flies were thick. The captain explained, "It's just like it always is. It's not fixed up for inspection at all."

The kitchen was next, and flies were about as numerous. Said the captain, "Those damned flies beat us back here."

The first year a second camp was scheduled to go in on the Coeur d'Alene River near the mouth of Big Creek. A major drew this tent camp. The camp already there was an all-lumber camp, the first to go in, and was completed, or nearly so, with Lieutenant Hazeltine in charge. The latter's company consisted of Idaho enrollees. The major appeared at Hazeltine's camp and told him he wanted Hazeltine's saws and hammers. The lieutenant said that he wasn't through with them and couldn't let him have them. The major insisted, saying his men were sleeping on the ground, not even any canvas over them, and the lieutenant's men had closed-in, heated buildings. The deadlock was only broken by an offer of Forest Service tools until the major could get army equipment of his own. Rank didn't count in this case because each was an independent commander reporting directly to Fort George Wright.

By the beginning of the second year, educational advisors were provided for each camp. They were civilians with previous training. A room or space was set up for a library and a supply of books arrived. Enrollees were encouraged to read and study. In a few of the camps many of the boys first had to be taught to write their names, for each enrollee was required to sign the payroll. They learned to read and write and do simple arithmetic. This was a sad commentary on the educational program of some of the states. The educational advisors took what they found and served a most useful purpose in improving the knowledge and ability of the enrollees. Schoolwork was not compulsory and they had to be diplomats.

One night I received a call from the local hospital saying a Forest Service man had been in an accident. I found William Nagle on the operating table. He was the regional liaison man between U.S.F.S. and the army in the CCC program. His car had collided with another in the fog. The doctor was sewing up deep cuts on his nose and both upper and lower lips. Bill was pretty groggy from the anesthetic but he knew me. The doctor said he'd be OK but would have some scars. Bill said weakly, "That - won't - hurt - me - for - hard - work. I - wasn't - figuring - on - going - to - Hollywood - anyway."

On one occasion a lad came to my home on a weekend and wanted to talk with me. He said one of the boys at camp had told him that the Forest Service would pay one dollar each to have "squeak trees" located and cut down because they caused forest fires. He figured he could make some extra money on Saturdays or odd times, and he wanted to get in on the ground floor. After visiting with the chap a few minutes—asking him if he knew of many squeak trees not far from camp and whether he expected much competition for the work, I broke down and told him he'd been jobbed and proposed that he figure out a good way to turn the tables on his chum in camp.

I've heard of snipe hunts, and new hands sent for left-handed monkey wrenches, but this was a new one. One stunt was to find a boy fearful of wild animals. The others would build up fear for days with stories of horrible manglings by bears or cougars, or by some beast out of the imagination. Then some dark night they would get short candles and put them in cans close together. The fearful one would be enticed a short way into the woods when a tormenter would scream, "There's one of 'em now! Run!" He turned to point at the sinister bright shining eyes.

Games were figured out, especially if a money angle could be tied to it. Snake fights and rat fights were common. Some kinds of the little rodents fight to the death and, if a company could find one of demonstrated ability, they would bet what they had down to the last penny.

Some camps went strongly for boxing or baseball or some other more orthodox sport.

Division of the enrollees between the army and the technical agency could make difficulty. Instructions said that the army would retain twenty-six enrollees and turn the remainder over to the work agency. Needed were cooks, kitchen help, company clerk, library assistant, truck drivers for the army vehicles, infirmary assistants, night watchman, maintenance workers, laborers for new walks or fences, or garbage disposal and trainees learning some of these jobs. Some jobs went on seven days a week while men were not required to work more than five days. Men otherwise turned out for project work were kept in camp part time as substitute headquarters men. If total enrollment went down due to men going "over the hill" or completing their enrollment, the project crews suffered the loss so the camp jobs could go on.

Some camps had more difficulty than others. Many of the commanders were most cooperative and found ways of holding their needs down to further the work projects. Some used camp jobs as disciplinary measures instead of confining the boys to quarters. Some camp work was assigned to "goldbrickers" who were infirm until the medic made his round.

The two-hundred-man camp was a suitable size so far as library, medical, and housing features were concerned, but it did not fit well into the project work. The Forest Service had always operated on the basis of getting a camp just as close to the work as possible. Even on forest fires this was practiced. A twenty-five-man crew (or smaller) saved a lot of travel time. With the CCC's—trucks were resorted to and all men on project work were hauled back and forth. Stake trucks with bench seats and bows and canvas covers were used.

It was rumored that at the inception of the CCC pro-

Courtesy U.S. Forest Service
A CCC BOY BUILDING A FIRE ROAD WITH A D-4 DOZER
ON THE COEUR D'ALENE FOREST
Many enrollees became skilled tractor operators and later qualified for employment as such

gram the Chief Forester in Washington, D.C., R. Y. Stuart, tried to convince President Roosevelt that smaller than two-hundred-man units were more practicable but was told, "Mr. Forester, the camps are to be in 200 man units and if you can't handle the work that way another Forester can."

It worked out that "side" or "spike" camps were allowed in special instances where urgent work was remote from the base camp. This course was avoided when possible for it meant extra travel for the commander and superintendent. In some cases a paid cook was financed from Forest Service funds.

The overhead in the CCC camps were not under Civil Service appointment. The technical agency selected them and picked the best qualified for each job. Project superintendents were carefully chosen from among Forest Service assistant rangers, road superintendents, blister rust unit supervisors, lumber company foremen, or otherwise experienced men. Crew foremen were drawn from among those experienced in woods work and accustomed to directing men. The men had to be on one of the Congressmen's or Senator's appointment lists but anyone could get on the list by writing a letter to Washington, D.C. There were so many more names on the lists than there were jobs that men could be selected entirely on their merits.

During my employment in the Forest Service I served under four Republican and three Democratic administrations and this was the closest approach to political patronage that I ever encountered.

An allotment was made to the technical agency of a certain amount a month per camp. The lineup of projects was influenced by the funds available. With many of our camps we determined on one road crew, the rest of the manpower to be used on blister rust. This was hand work and required little expense except for foremen's and checkers' salaries, and truck operation. The remainder of the allotment was available for bulldozers, graders, compressors, gasoline, dynamite,

Courtesy U.S. Forest Service

A CCC CREW MAKING A FIRE LINE ON THE BOISE NATIONAL FOREST, IDAHO

They learned to use hand tools and to work as a team

culvert material or whatever was needed for the road building, including the road foreman. By that practice we were able to use heavily of manpower on blister rust and heavily on cash on road construction.

The CCC program was splendid for hundreds of thousands of young men. Many received much-needed schooling. All had the benefit of travel. They learned discipline and supervision. They learned to get along with others. Undoubtedly many of these fellows, for they were just the right age, got into the military in World War II and their hitch in the CCC's made them better soldiers and cut down training time.

They learned to work. Many, from the big cities, had never done a day's work in their lives. Many first-class truck drivers were developed. At one of the colored camps a heavy road job was under way. Some well-qualified bulldozer men were produced, as well as compressor and jackhammer men. About all the crews were used on fire-fighting jobs, for which they were given training. This involved the use of small tools, axe, shovel, saw, and pumps. They had first-aid training and used it. Some did carpenter work and others helped develop camp and picnic grounds.

On our forest our largest single activity was eradication of currant and gooseberry bushes. The blister rust threat was the justification for our getting such a large allocation of camps. I doubt if any forest anywhere had as many camps as the Coeur d'Alene. This work was tedious and had no great appeal to the boys, yet it did teach and require the virtues of observation, thoroughness, and stick-to-itiveness. All of the foremen helped the educational advisor by conducting evening classes on subjects they were expert on. One foreman had been covering some geology features of the territory and described the action of glaciers in preceding ages. One enrollee asked, "Well, where are the glaciers now?" The foreman replied, "Oh, they have gone back for more rock."

Some of the boys became pretty fair cooks, others learned

to type. Unquestionably quite a few lads were diverted from a life of crime. Some of them were from the slums of big cities and were well started in that direction. One commander (and he was a good one) shook down his company once a week and confiscated razors, knives, guns and lesser weapons.

The boys came from every state and of course their backgrounds varied widely. Jackman tells of one camp he visited in Sherman County, Oregon, the only county out of the state's thirty-six that does not have timber. The camp worked under the Soil Conservation Service and not a tree was in sight. Jack had been attracted by a fine little camp newspaper and wanted to meet the editor. He was an enrollee from Boston and this conversation followed.

JACKMAN: "What did you do when you found you were to go to Oregon?"

ENROLLEE: "I went to the library and studied up on Oregon. I found it had about a third of the nation's timber, so I studied up on timber and then they sent me here!"

JACKMAN (attempting consolation): "Oh, well, this isn't much like Boston."

ENROLLEE: "Hell, mister! This isn't even much like Oregon!"

No one can say what the schooling and work habits were worth to the boys. That cannot be measured in dollars and cents. But it paid handsome dividends to the nation.

A new Youth Corps or Job Corps is getting under way in 1965 as a successor to the CCC program of the thirties, but with certain desirable changes.

1. 100-man camps (some, at least).
2. Provide hired qualified cooks.
3. Do away with divided authority and responsibility.
4. Enlarge the training program.
5. Small allowance monthly. Withhold most of pay until completion of enrollment.

CHAPTER THIRTEEN

FOUR FORESTS

All Four

THIS CHAPTER CONCERNS THE CARIBOU, LOLO, COEUR d'Alene, and Whitman National Forests.* Much of this book pertains to the activities and episodes on these forests, but some things are not a part of the forest activities chapters.

These forests are located in three different administrative regions. The Caribou is in the Intermountain Region, the Lolo and Coeur d'Alene are in the Northern Rocky Mountain Region, and the Whitman is in the Pacific Northwest Region. They lie within four different states. The first is in extreme southeastern Idaho and southwestern Wyoming. The second is in western Montana, the third is in the Panhandle of northern Idaho, and the fourth is in eastern Oregon.

The Caribou produces more forage to the acre and carries more domestic livestock than most any other national forest. The Coeur d'Alene, for its size, is the outstanding white pine producing forest. The Whitman has a complete and well-balanced catalog of resources and activities. The four give an accurate cross section of a major segment of the national forest system.

I was Supervisor of each of the four units, covering a total span of well over three decades.

---

* For all national forests, see Appendix.

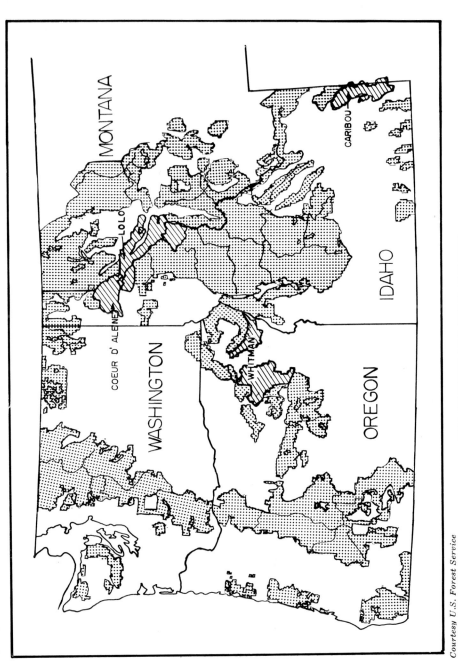

MONTANA

LOLO

COEUR D' ALENE

WASHINGTON

WHITMAN

OREGON

IDAHO

CARIBOU

Courtesy U.S. Forest Service

MAP SHOWING THE "FOUR FORESTS"

How they fit into the national forest picture of western Montana, Washington, Oregon, Idaho, and western Wyoming

Caribou Forest

The Caribou National Forest was established in 1907. The name does not come from caribou, the North American reindeer. No caribou ever grazed on the Caribou Forest since the coming of the white man. The name came from Caribou City, an early-day mining camp. On the old maps it was spelled "Cariboo City," as in the Cariboo in British Columbia. I think the Geographic Board, names authority in United States, has officially adopted "Caribou" as the correct spelling. Everything is "Caribou" now. There's a Caribou City, Mountain, Range, Basin, Guard Station, Forest, and County. The latter was named when the citizens of Bingham County decided they wanted two county seats instead of one. The Soda Springs people no longer have to go to Blackfoot, Idaho, to do their county business. They didn't realize, though, that the new county would hit a jackpot when the phosphate business developed. The Caribouers

Courtesy U.S. Forest Service
GREAT NUMBERS OF LIVESTOCK GRAZE ON THE
CARIBOU NATIONAL FOREST
Here are sheep on summer range

missed one bet. There is no creek bearing this name. The Caribous are on McCoy Creek.

Headquarters of the forest was formerly at Montpelier, Idaho, on the Old Oregon Trail. Now Pocatello is the Su-

Courtesy U.S. Forest Service
A PLEASANT ROAD THROUGH MARBLE CANYON
The way leads through a grove of quaking aspen on the Caribou National Forest, mostly in southeastern Idaho.

pervisor's home, three small divisions of the Cache having been transferred to the Caribou.

Montpelier is at an elevation of 5,963 feet and Star Valley, in Wyoming, on the east side of the old Caribou, is situated at 6,200 feet. It is a good hay and cattle valley, with plenty of snow. It gets frosty in the mornings, too. An elderly Star Valley lady decided to go to Southern California for the winter. By Christmas she wrote that she was so homesick for Star Valley she didn't know how she was going to stand it.

This forest is outstanding for its forage production. It is a sheepherder's paradise. Old-timers told us they once could see a herder's teepee on every knoll. In 1964, 120,677 sheep were summered on the old Caribou. This is as many sheep as are grazed on all the national forests of Oregon and Washington combined.

Prior to 1930 the recreational use was very light. The big game population was small, too. By 1964 the forest reported 328,500 recreational visits. This, more than anything, indicates the increase in population and the increase of the free time of individuals. An increase of roads and autos permitted more people to enjoy the mountains. The increase in big game numbers provides added incentive, too. The forest reported several thousand mule deer and a substantial elk population in 1964.

The forest claims the largest known phosphate deposits in the world. Four phosphate processing companies, all million-dollar industries, are operating on the forest. Phosphate leases cover over 28,000 acres.

The program of multiple use applies on the Caribou, even though the grazing of domestic livestock is by far the major use on the forest.

Lolo Forest

The Lolo National Forest in the 1920's extended from just west of Missoula, Montana, westerly to the Idaho line east of Wallace, Idaho. The Elk Summit and Powell Districts in Idaho, actually part of the Selway Forest, were administered with the Lolo. In later years consolidation and transfers were made, and changes in forest boundaries, but the name has stuck. It is tied in with Lolo Pass, Lolo Hot Springs, and Lolo Creek.

Lewis and Clark, in 1805, on their westward journey, went through Lolo Pass into what is now Idaho. Chief Joseph, on his historical eastern trip from the Nez Perce country, outmaneuvered the United States Army on Lolo Creek. Captain Charles C. Rawn, from Fort Missoula, was sent up Lolo Creek to intercept and take Chief Joseph and his Nez Perce people. The captain was accompanied by 4 officers, 25 enlisted men, 150 citizen volunteers, and 25 Flathead braves. The soldiers selected a spot to make a stand, cut

Courtesy U.S. Forest Service
A MILLION-DOLLAR PHOSPHATE FERTILIZER PROCESSING PLANT
ON THE CARIBOU

trees and laid up logs four or five feet high as a barricade. Behind this they would waylay the warriors.

The chief had scouts ahead. They discovered the soldiers and alerted Joseph. During the night he moved his entire party out of the canyon onto the ridge to the north, bypassed the army without being seen or heard, and was miles to the east before the maneuver became known. The redoubt, occupied from July 25 to 28, 1877, was named Fort Fizzle. All that remained in 1925 was a little rotten wood.

To a forester an assignment to the Lolo National Forest afforded unusual contacts with other foresters. The Regional Forester of the Northern Rocky Mountain Region and his assistants are there and so are headquarters of the United States Forest Experiment Station. The Montana State Forester is located in Missoula. The School of Forestry of the University of Montana gives contacts with students and faculty. It is the location of the Blackfoot Forest Protective Association. When I was there both the Missoula Forest and Lolo Forest headquarters were in Missoula. Monthly meetings of the Northern Rocky Mountain Section of the Society of American Foresters helped all of us. Missoula was a small city and was forest-oriented.

U.S. Highway 10 extends from Missoula to Lookout Pass on the Idaho line and is within or close to the Lolo Forest all the way. West of St. Regis it climbed for several miles to the top of the "Camel's Hump" only to descend an equally steep grade to the St. Regis River. During the fire season the ranger at St. Regis had a fireman located on the hump to give out fire prevention pamphlets and a friendly word of caution about fire. No one objected to stopping, in fact 99 percent wanted to stop and let the motor cool off. Most of the radiators were boiling. One fellow commented, "I think that hill would make a Franklin boil." Nowadays he'd probably say "Volkswagen."

I was only minutes behind a touring car that wrecked on the west side of the hump. A man and his two youngsters were en route to Spokane. A two-foot Douglas fir had fallen

Courtesy Idaho State Historical Society
CHIEF JOSEPH OF THE NEZ PERCES
He outmaneuvered the United States Army on Lolo Creek in 1877. See pages 283 and 284

across the road, resting on the outside shoulder and a high inside bank. The grade was steep and he was going too fast to stop. The radiator just barely slid below the tree trunk and everything projecting above that level was sheared off smooth. The man was decapitated.

The St. Regis Ranger Station was up the St. Regis River along the Wallace branch of the Northern Pacific. One used a hand speeder or walked the ties the two miles out from St. Regis. One winter day I walked out to the unoccupied station. As I got to the dwelling I was alerted by a noise at the storehouse fifty yards away. I walked to the other building and, hearing someone inside, called, asking who was there. Getting no response I spoke again, telling the person to come out the way he went in. Shortly a young transient crawled out the broken window. He had a Forest Service packsack and a few dollars' worth of government groceries. After a walk on the ties he was turned over to the sheriff.

Following the construction of a new station at St. Regis, a good highway was built down the St. Regis River eliminat-

Photo by Charles Simpson, Baker, Ore.
TOP OF THE CAMEL'S HUMP ON U.S. 10 IN 1930,
LOLO NATIONAL FOREST
Model T's and no blacktop—no snow removal

ing the Camel's Hump. No one regretted this. It belatedly provided good access to the abandoned ranger station, which had served its usefulness.

Usually thought of as upland prairie, Montana is a big state with plenty of room for variation. The part west of the Continental Divide is mountainous and timbered. The farther west one goes the more dense and bigger is the timber and the Lolo is at the extreme west end. Western larch and Douglas fir were the dominant species, with good ponderosa pine at the lower elevations and some western white pine higher up and near the Idaho line. Due to the density of timber, brush, and windfalls, usable range was scarce. Several bands of sheep were able to use some of the high country, heavily burned in 1910 and taken over by browse and weeds.

Photo by Idaho Department of Highways
U.S. 10 IN 1965. NOW INTERSTATE 90
The Camel's Hump was abandoned and a new canyon route substituted. This picture was taken west of the Camel's Hump on the Idaho side of the state line.

The sheep were shipped on the Northern Pacific from Yakima and central Washington points on a feeding in transit basis. Late in the summer fat lambs were reloaded on cars and shipped to Omaha. The ewes were trailed west to the wheat fields of eastern Washington and reached the Yakima country about Christmas.

White-tailed deer were numerous and provided big game hunting. Many elk and some moose were on the East Selway.

Lightning storms were frequent and severe. July and August were usually rainless. Lots of fuel was everywhere. The combination created a high fire hazard and made the fire protection job important.

Two sizable sawmills and some smaller ones were sawing government timber and the easily accessible logs were in reasonable demand.

We had scarcely thought of sustained yield in the twenties. We had enough for all comers, so everyone believed.

Photo by Bill Bell, Missoula, Mont.

PACKING OVER LOLO PASS TO THE LOCHSA RIVER IN THE
SPRING OF 1923

Even then Lewis and Clark's route was no highway. Today a broad blacktopped road
traverses the region.

Coeur d'Alene Forest

The name Coeur d'Alene is of French origin, meaning
"Heart of an awl," or "Sharp Hearted." The Coeur d'Alene

Photo by Idaho Department of Highways
LEWIS AND CLARK HIGHWAY ALONG THE LOCHSA RIVER, 1965
"If Sacajawea were alive today she would turn over in her grave," says Tony Branden-
theler, of Baker, Oregon.

Indians bore the reputation of being sharp or keen in trading, hence this name was used by the French.

The beautiful timber-lined lake separating the mountain country from the valley early took the name Coeur d'Alene. It is twenty-two miles long, averages two and a half miles wide, and its deep bays and long fingers create a shoreline more than one hundred miles long. It laps at the edge of the Coeur d'Alene National Forest and provides a sandy bathing beach right in the town of Coeur d'Alene.

Hayden Lake, a few miles from Coeur d'Alene, is smaller but is truly beautiful. Here are the Country Club and golf course, and many summer homes of Spokane people as well as those of local residents.

The Coeur d'Alene National Forest was established November 6, 1906, with headquarters at Wallace, Idaho. Shortly it was divided, the south portion becoming the St. Joe National Forest, with headquarters at St. Maries. Headquarters of the remaining part was moved from Wallace to Coeur d'Alene. It is one of the smaller forests, with 723,436 acres of government land. Yet after the early-day split no changes were made until recently when the St. Regis District, formerly on the Lolo in Montana, was put under the administration of the Coeur d'Alene officers.

It is a timber producer of the first order. It produces western white pine, the choicest and most valuable of all western woods. The climate suits white pine. Moisture is ample and soil is satisfactory. Ridges and mountaintops are almost entirely soil-covered. Rock outcrops are rare. Trees grow large and dense right over the ridges. Prevalent winds are from the southwest and for centuries they have picked up dust from the rich prairie lands of Idaho and east-central Washington and deposited it in the mountains. The process has accelerated with vast summer-fallow fields.

The forest boasts no Mount Hoods, no Mount Jeffersons or Eagle Caps. It has no wildernesses or high mountain lakes. Nature evidently was satisfied with the big lakes adjoining to the west. There are no large meadows, and no large open plateaus or grass-covered ridges. Domestic livestock use is insignificant. Except on the burned timber areas even wild game do not find luxury living.

THE MULLAN TREE
A white pine blazed by Captain John Mullan on July 4, 1861, in an area named Fourth of July Canyon. Coeur d'Alene National Forest. See page 293.

Multiple use, yes, but the five-pointed symbol has a big bulge where timber production shows. It is severely lop-sided. The forest is a forester's and a lumberman's pride and joy. Over 125 million board feet of timber products are marketed each year.

Lewis and Clark managed to miss the Coeur d'Alene on their westward journey but Captain Mullan made history from 1859 to 1862 when he was building a military road from Fort Benton on the Missouri to Fort Walla Walla near the Columbia. He laid out 624 miles of wagon road at a cost of only $230,000. On July 4, 1861, he camped on a small tributary of the Coeur d'Alene River and the area became known as Fourth of July Canyon. On a white pine tree about forty inches in diameter he made a rectangular

Photo by Idaho Department of Highways
FOURTH OF JULY CANYON IN 1965
The former U.S. Highway 10 is seen at the left. Four roads and 104 years after Captain Mullan slept here.

blaze and carved the date "July 4, 1861." This became
known as the Mullan Tree. Many years later it was protected by an iron fence, for it was only a stone's throw from

Courtesy U.S. Forest Service
WESTERN WHITE PINE—THE QUEEN OF THE FOREST
Area on Deception Creek, Coeur d'Alene National Forest, Idaho

U.S. Highway 10. The tree was broken off about twenty feet from the ground on November 19, 1962, in the storm that created such havoc on the West Coast. The three-hundred-year-old stump has been preserved.

A few miles west of the Mullan Tree was an excellent illustration of history and the march of civilization. The brush-screened scar of the Mullan wagon road showed where the captain had taken the easy and quick route to build his road. This had given way to a county road, built more with an eye for easy use rather than easy building. Then U.S. 10 showed the influence of engineers with their love for tangents and their disregard for dollars. From the opposite hill all three roads could be seen. Now the route is known as U.S. Interstate Highway No. 90 and a double highway with wide

LOOKOUT PASS, ON THE IDAHO-MONTANA DIVIDE
Burned in 1910, the area is pictured as it appeared on February 26, 1939. There are skiers on their own going both up and down, scattered clumps of young trees, and a gravel road.

separation is in use. The U.S. 10 engineers were pikers. Time marches on!

The "Coeur d'Alenes" refers to the important mining country in Shoshone County, Idaho, along U.S. 10 and within the Coeur d'Alene National Forest. Wallace, Kellogg, Mullan, and Burke are all supported by mining. Silver, lead, and zinc are the principal metals but some gold and other values show up. A gigantic fault or earth slip has been traced through the region and the formation on one side of the fault exactly matches that found twelve miles away on the opposite side of the fault. Geologists call this a "thrust fault" to show the difference from those formed by straight up and down movements. Most of the ore is found deep underground, although the Bunker Hill was alleged to have

Courtesy U.S. Forest Service
LOOKOUT PASS, AS IT WAS IN 1961
The trees are thicker and larger, the ski lodge and lift attract many skiers, and the wide, blacktop highway—Interstate 90—now provides a ready means of access to the area.

THERE ARE IMPORTANT STANDS OF PONDEROSA PINE
IN THE WHITMAN NATIONAL FOREST OF OREGON
Ranger Robert Harper is not actually marking this unusual tree for cutting

been discovered by a prospector's mule pawing on the surface.

The Bunker Hill and Sullivan Mining and Concentrating Company has a large smelter at Kellogg where it smelts its own ore and does custom smelting for other companies. It has paid good dividends for years. The Sunshine Mine, west and south of Kellogg, is reported to be the largest producer of silver in the world. The Morning, Golconda, and Hecla are well-known mines in the area. There are others.

Whitman Forest

This forest was named after Marcus Whitman, famous missionary who came West before the wagon trains traveled the Oregon Trail. He located in the Walla Walla country

Photo by Jack Eng, Baker, Ore.

A GOLD DREDGE FLOATING IN A POND OF ITS OWN MAKING
The dredge worked three shifts daily in Sumpter Valley, west of Baker, Oregon, and its huge dippers handled thousands of yards of gravel each day.

and was massacred by Indians in a local uprising. The Old Oregon Trail crossed Baker Valley and a portion of the present Whitman National Forest.

The forest headquarters was first located at the mining village of Sumpter, Oregon, but in 1920 it was moved thirty miles to Baker. About the same time the Minam Forest, located north and east of Baker, was annexed to the Whitman. A District Ranger was stationed at Sumpter for several years after the Supervisor moved out. Then he, too, moved to Baker.

By 1952 conditions justified a more drastic change. The Supervisor's office at Enterprise was closed and the Wallowa Forest with four ranger districts was put under direction of the Baker supervisor. Blue Mountain, one of the six districts on the Whitman, was assigned to the Malheur at John Day. Changes are justified by more and better highways, the use of radio communication, and more administrative responsibility placed on District Rangers. In the Blue Mountain case, a change in lumber company ownership and operation entered in.

The Supervisor job of the Whitman has been a stepping-stone for more than the usual number of men. Robert M. Evans became Regional Forester of the Northeastern U.S. Region. Walt Dutton moved on to Assistant Forester in charge of Grazing, Washington, D.C. John Kuhns became Assistant Regional Forester in Portland, Oregon, heading Public Relations. Lester Moncrief was promoted to Assistant Regional Forester in charge of Personnel in Ogden, Utah. Harold Coon, after another assignment, also went to the Intermountain Region as Assistant Regional Forester in charge of Fire. Jack Smith is now in Juneau, Alaska, Assistant Regional Forester handling Timber Management.

The Whitman has some of everything, while many forests, as shown, run heavily to some one resource. The Whitman is a good training ground.

The Whitman has been in the timber business since early days. There are yellow pine, Douglas fir, western larch, white

fir, lodgepole pine, spruce, juniper, and even a few pockets
of white pine. There is range management with both sheep
and cattle. Deer and elk are plentiful. Lakes and streams
and storage reservoirs provide much needed water for irrigation, power, and domestic use. There are the Eagle Cap
Wilderness Area, Anthony Lakes ski area, and lesser recreation uses abound because of accessibility. Mining at one
time was important and may be again. One of the recreational uses is to search out the old mines that turned out
millions.

This is a mineralized area and gold hunters settled the
country. Auburn, where gold was discovered, had a population of five thousand in Civil War days. The only rem-
nant to be found today is the graveyard with about twenty-
five markers. Two stones side by side recall two young men
who died the same day without "due process of law." In

Courtesy U.S. Forest Service
UPSIDE-DOWN LAND
Gold is most often found on bedrock so the dredges go deep. They leave a gravel waste
and ruin beautiful mountain valleys.

1940 twelve gold dredges operated within twenty-five miles of Sumpter. By 1955 all had been shut down and dismantled. The fixed price of gold and the rising cost of operating them made the work prohibitive, and during World War II gold mining was classed as nonessential.

A few of the gold-in-place producers were Cornucopia, Buffalo, Cougar, Monumental, and Red Boy mines. A gold exhibit in the U.S. National Bank in Baker interests visitors. A seven-pound nugget found in 1913 is valued at $2,500. Many smaller finds include nuggets of various sizes, dust, and thread or wire gold. Rich ores from tunnel or shaft are also on display.

The Whitman, all in all, is diversified and presents a most interesting multiple-use enterprise.

MAIDA VICTOR CALIPER SCALING IN THE WOODS
She replaced a man in service during World War II. Whitman National Forest (Oregon) timber sale.

CHAPTER FOURTEEN

ROCKY MOUNTAIN SPOTTED FEVER

A Night Ride

"COME OVER AND HAVE A LOOK AT THIS MAN'S BACK AND you'll never forget what spotted fever looks like." It was Doctor Hobson, a Missoula doctor speaking, and the patient was Charlie Johnson, a Forest Service trail foreman. He was lying on a steel cot at the Elk Summit Ranger Station on the Idaho side of the Bitterroot Divide. His skin was a fiery red with a fine network of veins standing out even redder than his skin. Once in a great while one sees a man whose cheeks are red with noticeable veins. This was far more vivid.

The doctor was fishing along the Lochsa River when word came that the foreman was sick and had been brought to Elk Summit from the trail camp. The doctor was located and agreed to make the trip. We left Powell Ranger Station, the end of the road, with saddle horses about 10:00 P.M. The trail was good but through dense timber and there was no moon. It was pitch dark. It was uphill all the way but we kept moving at a good lively walk. We started with three lights, using one at a time. The last one failed just before dawn. We covered the twenty miles before sunup.

The doctor at once examined the patient and gave him such relief as he could with the limited supplies available. The man was delirious but quiet. His activity consisted only of moving an arm slowly, passing his hand across his face as though brushing off an imaginary fly. He died about 10:00 A.M. as we watched him helplessly.

The following morning we prepared to return the way we had come, but with the corpse. Have you ever packed the cold body of a man you knew, over the back of a pack mule —feet hanging down one side, head down the other? First he was wrapped in two blankets, then in canvas. Ropes were tied tightly to keep the coverings in place. He was placed stomach down over the Decker packsaddle with a sling rope on either side binding him firmly.

Prior to 1906 several theories attempted to account for Rocky Mountain spotted fever. For years it had been common, particularly in two areas in Montana—the Bitterroot Valley and adjacent territory and in Carbon County east of the Continental Divide. It is not confined to Montana, or is it confined to the Rocky Mountain States. It has occurred throughout most of the United States. About five thousand cases have been reported. At first more than one out of six proved fatal. The fatality rate was much higher in the two areas. Indians frequently died from it, and early settlers in the Bitterroot country were especially hard hit. I knew of one sheepherder on the Lolo Forest who was found dead in his tent, his dog by his side, a spotted fever victim.

Indians believed that the water from certain sources was the cause. The belief developed that the east side of the Bitterroot Valley was free of the disease but the west side was not.

Wood Ticks

In 1906 two scientists produced Rocky Mountain spotted fever in guinea pigs by allowing infected wood ticks to bite them. First thought was that the ticks responsible were a special kind of tick and they were named "spotted fever tick." Later studies proved that the spotted fever tick is the common wood tick of the Rocky Mountains. Not every wood tick can cause spotted fever. Only those infected by the virus can transmit the disease.

In addition to the Rocky Mountain wood tick there are

two other ticks in the United States that are carriers of the disease to man. The American dog tick is distributed in the coastal areas of California and throughout the Central and Eastern States. The lone star tick is found in the South-eastern and South Central States. The Rocky Mountain wood tick is distributed in the thirteen Western States except for the coastal areas and the southern parts of the southern tier of those states.

The Rocky Mountain wood tick develops through four stages.

1. Eggs are laid in a pile on the ground, among leaves, dead grass, or sticks. Each female lays several thousand eggs.

2. The eggs hatch after thirty-five days, the same summer as laid. The six-legged larvae are ready to feed on their chosen hosts. They crawl up grass or weeds and become attached to passing animals such as squirrels or other rodents. When full of blood they drop to the ground, become inactive, and go through a change.

3. The second-stage tick or nymph has eight legs and is larger than the larvae. The nymphae usually hibernate over the winter and come out the next spring. They also seek the blood of small rodents. After feeding for a week they drop off, hide for three weeks.

4. They emerge as adult ticks—male and female. Usually this is too late in the season or too hot and dry for them to find hosts and they hide near the surface of the ground until the following early spring. Then they crawl up on grass and bushes and get onto passing animals. Their hosts are primarily the larger animals, cattle and horses, sheep, and deer and elk. The blood of humans is very acceptable, for blood is what they are seeking. After the female is much enlarged from the intake of blood she drops to the ground. In about a week she begins producing eggs at the rate of about three hundred per day. This is normally continued for three weeks.

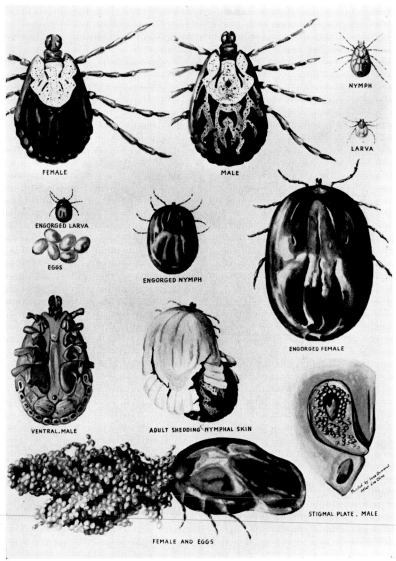

FEMALE

MALE

NYMPH

LARVA

ENGORGED LARVA

EGGS

ENGORGED NYMPH

ENGORGED FEMALE

VENTRAL, MALE

ADULT SHEDDING NYMPHAL SKIN

STIGMAL PLATE . MALE

FEMALE AND EGGS

Photo by U.S. Public Health Service, Hamilton, Mont.

LIFE CYCLE OF THE WOOD TICK

A female wood tick lays about six thousand eggs. No wonder ticks are plentiful. Adults
are enlarged about nine times.

Spotted Fever Laboratory

The United States Public Health Service established the Rocky Mountain Spotted Fever Laboratory at Hamilton, Montana. Workers there studied the life history of ticks, made a determination of host animals, traced the transmission of spotted fever virus from ticks to man, and finally developed, in 1925, a serum known as the Spencer-Parker vaccine. Men working at the laboratory risked their lives daily in experimenting with the ticks. Five of the workers died from accidental infection. All of the laboratory cases prior to the vaccine were fatal. Of nine cases of spotted fever among laboratory workers since vaccination was initiated only one proved fatal.

The vaccine was developed by permitting infected ticks to bite horses, then withdrawing blood from the horses. This was refined, diluted, treated, and finally injected into man's bloodstream to build up immunity. Guinea pigs and other animals were tested first, then the doctors and other laboratory workers tried out the vaccine on themselves, subsequently allowing themselves to be bitten by infected ticks. Present vaccine is made from egg material and should not be given to persons allergic to egg protein.

When the vaccine became available, the Forest Service field people took it, as did most of the loggers, stockmen, and ranchers whose work required them to go into the woods and fields. At first three shots were given a week apart. Then it was changed to two vaccinations. It actually results in a very mild case of spotted fever. The vaccine at first was supplied free and the Public Health nurse made the inoculations, or one could go to his own doctor. Doctors now get it through druggists. Normally there are no bad aftereffects but the skin on the arm becomes slightly red and an itching sensation may follow. Some persons have really sore arms for a few days.

One spring following my first shot my arm swelled alarmingly and itched unbearably. When I went back for the second vaccination I accused the nurse of giving me the dregs from the bottom of the bottle. She explained that the vaccine was uniform and told me after that shot to take a

Photo by U.S. Public Health Service, Hamilton, Mont.

A WOOD TICK CLIMBS UP A BUSH OR A SHRUB AND HANGS
BY A REAR PAIR OF LEGS WITH ITS HEAD DOWN

Waving its front legs, it grabs hold of the first moving object it contacts and climbs aboard. Beware of its bite! Adults are a quarter of an inch long and these ticks are enlarged about ten times.

good dose of laxative and I would have no trouble. Sure enough, no disagreeable effects followed. Even at the worst the cure (preventative) is far better than the disease.

Avoiding Tick Bites

The surest way not to get tick bites is to stay inside. A forester, sheepherder, or log cutter can hardly do that. It helps to wear high-topped shoes and long, heavy socks with trouser legs tucked inside the socks. The ticks travel upward on a host and thus stay on the outside of the clothing longer. Gals should wear breeches with closed ankles or they are really vulnerable. Wood ticks are the original hitchhikers. They climb a bush or shrub, then hang by their hind legs, head down, waving their front legs (thumblike) in the air. The first living object they contact becomes their host. They swing on as a rider of the rods would swing onto a freight car.

It's worthwhile to take a five now and then and pick off all the ticks one can find on clothing. They are more numerous in early spring. On a short day's hike I once picked off more than ninety ticks. The infection rate among ticks varies from less than 1 percent to 5 percent in different localities and in different seasons, and may be as high as 11 percent in some areas. It is impossible to tell whether or not a tick is infected by looking at it. There may not be an infected tick in a hatful, and then again the first one to bite might be a bad one.

Don't ever wear your hair Beatle style. If your hair is clipped short around the neck you are almost sure to feel the crawlers and easily find them. When a tick bites he exudes an anesthetic so his incision is painless. I have seen men remove ticks that had been attached long enough to fill with blood and resemble a blueberry. It pays to remove all clothing at least at noon and night. Search your body and clothing carefully.

If you find a tick don't try to "unscrew" him. They

don't bite that way. An application of turpentine is sometimes effective. It is better to pull the tick loose with a
steady pull, using the thumb and finger to hold the tick. An
effective method is to take a piece of cardboard and make
a hole through it the diameter of a lead pencil. Hold the
cardboard against your flesh with the hole over the attached
tick. Light a match and apply the flame to the hole in the
cardboard. The tick lets all holds go and curls up in a hurry.
The cardboard merely protects your skin. Paint the point
of incision with a disinfectant.

A sheepman friend named Hale from Yakima, Washington, was in the Bitterroot Valley buying lambs. He had
been out visiting local herds and returned to Hamilton for
the night. He knew the spotted fever problem. He stripped

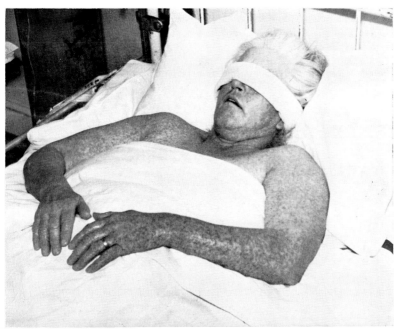

Photo by U.S. Public Health Service, Hamilton, Mont.
THIS PATIENT HAD NOT BEEN VACCINATED
The red-purplish rash of Rocky Mountain spotted fever. Cases are rare now because
vaccination is a sure preventative.

off and looked and felt himself over. He found a tick between his shoulder blades. He hadn't come prepared so he dressed and went to the drugstore and purchased a bottle of turpentine. Back at the hotel, he removed his upper clothes and applied the turpentine. It didn't make the tick let go immediately so he waited a few minutes, then made a second application. Still no results. Next he pulled gently but it didn't come loose. He had heard that ticks screwed their mouth parts into one's flesh. Although it was awkward to get the proper hold and his arms were almost paralyzed, he tried unscrewing the embedded tick. Still it didn't let go. Then it occurred to him maybe ticks were left handed and he was turning the wrong way. So he made another awkward try, this time turning to the left. He gave up, put on his clothes, and hunted up a doctor's office. He explained his difficulty and all his efforts. Then for the third time he removed his clothes. The doctor took a good look and exploded. "Hell, man, that isn't a tick, it's a wart!"

CHAPTER FIFTEEN

BURROS, BUCKBOARDS, BEDROLLS AND BEANS

Wrong Way Corrigan

THE NATIONAL FOREST'S "ANNUAL STATISTICAL REPORT" was a sheet about the size of a saddle blanket and included a multitude of figures telling all—so far as the doings of the year were concerned. The only machine on the Minidoka Forest in 1914 that would handle the form was an old long-carriage Oliver. It had a nonstandard keyboard. Hensley G. Harris was the forest clerk and the only clerical worker in the office. He typed with an original sort of two-fingered (each hand) touch system. Finally, after typing intermittently for two or three days, he rolled the finished product from the machine. The carbon paper was wrong side to. His language was inadequate.

Statutory Salaries

We fuss about centralized government and the control Congress has on many matters, yet in some respects there is less control of detail than fifty years ago. Up until 1920 Congress specified not only the salaries of Forest Supervisors but how many at each rate.

The Intermountain Region had the following yearly salaries, known as the statutory roll:

1—$1,600	8—$2,000
12—$1,800	1—$2,200

When I went to the Caribou in 1917 I drew the lone $1,600 supervisorship, but after one year a resignation occurred somewhere and I went up to one of the $1,800 spots. And did we envy the top dog who held that lone $2,200 place!

Rocky Mountain Canaries

Much of the trail and telephone maintenance was done with a two- or three-man crew. It didn't warrant a packer or camp mover. We used one camp several days to a week, walked to work several miles in the morning and back to camp after work.

Everyone has known or heard of a prospector who had a trained burro that hung around the camp or cabin, ate hot cakes or other leftovers and followed the prospector wherever he went. He was available to pack groceries, water, firewood, or whatever needed to be packed. He was the prospector's tried and true pal.

Someone argued that burros were the answer for moving camps used by small crews going beyond the reach of roads. They could be loaded after breakfast and taken right along with the trail men or the telephone repair crew. At night the beds and groceries, frying pan and coffeepot, would be right there. The crew could start fresh the next morning, right at the kitchen door. Basque sheepherders worked that way. The loaded burros grazed or trailed along with the sheep.

The Northern Rocky Mountain Region in 1926 shipped 150 burros from the Southwest to Missoula, Montana. They were all ages, all sizes, both sexes, mostly unbroken. They were parceled out to the forests and on to the ranger districts. Not all crews were fortunate enough to get some of Uncle's little helpers.

Halters and packsaddles were provided and the little beasts of burden were put to work. Anyway, it was tried. If the loads were a little heavy or the burros got tired, they would lie down. They got up when ready. At best they would travel one and a half miles an hour and the average trail worker would cover three and a half miles. Bred for generations on the desert, the burros weren't accustomed to walking through water, and most forest trails frequently ford creeks. Regularly the burros refused to set foot in the water. Even though small, the animals were too big for one man to carry across a creek. They couldn't be pulled or pushed. The trail worker who escorted the burros was often late getting to the place where his pardners quit work for the day.

An even bigger problem was that they would head for somewhere else when turned loose to graze. Somehow they had a faster gear on these occasions. It was not always pos-

Courtesy Charles Simpson, Baker, Ore.

MOUNTAIN CANARIES

Famous beasts of burden since biblical days. But no one loves them except their offspring

sible to catch up with them and get back to the job the same
day.

The venture might have been successful had we been
using prospectors or Basque sheepherders. But our Idaho-
Montana forest workers and the New Mexico burros failed
to become close pals and the burro project died on the vine.

Little Red Mule

On a pack trip on one of the forests we had a little red
mare mule. Horses are never supposed to be red. They are
sorrel or bay, but never red. The little mule was really a
rusty red. She weighed only about four hundred pounds
and she was too short-legged to walk as fast as the other
stock. So we turned her loose and when she got behind she'd
trot until she caught up. Because of her size she drew the
bedrolls, which made a bulky but not a heavy pack. On
occasion she'd grab a few bites of grass and might not stay
right on the trail.

I looked back at her once just in time to see a demonstration
of mule IQ. She had started to go between two big trees and
was stuck. She turned her head way around and looked over
the situation and, without a moment's hesitation, braced
her feet and gave a big shove backward. She came unstuck,
straightened up, walked around one of the trees, and came
on up the trail. Even lots of humans don't know how to get
out of difficult situations.

Where There's a Will

The Caribou Forest set some kind of a record in ranger
dwelling construction in the winter of 1918. The statutory
building limitation set by Congress was $650. Ranger Jim
Bruce was renting a little unpainted shack in Auburn, Wyoming, and he had secured a gift to Uncle Sam of some vacant ground there. We spent the entire allotment, the whole
$650, for material, and secured through channels the Sec-

retary's approval to build the dwelling. Approval was a requirement.

The crew assembled bright and early one Monday morning in December. The entire forest personnel was included except Miss Theresa Standing, the office crew. Rangers Jim Bruce, Auburn, Wyoming; Charles Spackman, Freedom; Camas Nelson, Snake River Ranger Station; Lewis Mathews, Grays Lake, Idaho; Bob Gordon, Georgetown, Idaho; and George Henderson and I from the Supervisor's office made up the seven-man brigade. Henderson and Spackman were skilled help, cutting rafters, hanging doors, fitting windows and supervising the rest of us who packed lumber, shingled, and pounded nails. We worked from frosty morning until too dark to see. There was no featherbedding. By Saturday night the house was ready to move into. The concrete foundation was poured before our arrival. True, there was no fireplace or patio. But it had two bedrooms and most of the essentials for living. Neighbors were most complimentary. It was the swiftest house building ever seen in Star Valley.

Jim Bruce's wife cheerfully cooked three meals a day for the hungry mob, doing it on what we knew as "contributed time." And she was pregnant at the time, though no one realized it. The groceries came out of the same fund as the cook's wages.

Model T Pioneers the Way

Clinton G. Smith, Chief of Silviculture, Ogden District Office, was inspecting the Minidoka National Forest the summer of 1916. He had a new Model T Ford touring car, personal property (government cars were unknown). Mileage of three cents was paid on official trips. After finishing his work on the Cassia–West Division, he decided to return to Oakley by way of Bostetter Ranger Station. There was a sort of wagon trail over the mountain, but putting a car over it was undreamed of. We thought him a bit loco. Fred

Betts, ranger on the district, and I, with ropes on our saddle horns, gave him a double tow from the Frazer Ranger Station to the top of "Deadline Ridge." This was about four miles and seemed three times that far. From there it was partly downhill and by removing some rocks, filling gullies with rotten logs, and using a shovel, he worked his way alone and made it through to Oakley. On the last lap he knocked out the plug in the crankcase and lost most of his oil. That was the first car to traverse the Cassia Division.

A Ranger Works His Plans

Sam "Buck" Buchanan was the District Ranger in 1914 and 1915 on the Rogerson District of the Minidoka Forest. His sole transportation consisted of two ponies he could drive to buckboard or ride and pack. His station was about midway of the district north and south and fourteen miles from mail and supplies at Rogerson. As things came up—word from the Supervisor's office, applications for small timber sales, scaling at the mill, survey jobs, meetings with stockmen on range or what not—it would go into his notebook under (1) North trip, (2) South trip, or (3) Town trip. When taking the North trip he'd get together everything he'd need—marking axe, scale rule, salt plan, compass, chain, blank sale contract and needed records, bedroll, and grub box. He'd pack and saddle up and strike out on a route to the north end to get where and when he was scheduled. When ready for town trip, he reviewed his notebook—loaded typewriter, needed forms, data, and files in the buckboard and away he went for Rogerson. Friends of his ran a hardware store and Buck engaged a store corner as a one-day office. He read his mail, prepared answers, made out reports, and added some more entries in the trip book. Then he headed back for his mountain station and prepared for a South trip. In spite of more elaborate work plans in use at times, Buck provided the best example of progressive travel and efficient use of time I ever knew.

Shanghaiing a Cook

Available cooks around most of the small forest communities were usually rare. So we put up with a cook's behavior that would have been intolerable with other members of the crew. The crew would do anything to pacify the cook in order to avoid taking over the cooking themselves.

One spring we were loading out a planting crew—groceries, equipment, beds, and tentage on lumber wagons. When ready to start the cook was missing. We found him in one of the saloons dead drunk. With main strength we got him on top of one of the loaded wagons. We tied him down hand and foot. We couldn't afford to let him fall off and get broken. He slept for several hours, the wagon serving as a cradle. Finally he awoke and demanded to be untied. When we reached camp he was sober enough.

After all, he was the life of the crew. One of his frequent instructions was, "Eat 'er up or you'll get 'er in the soup tomorrow." And he meant it, whether it was rice pudding or oatmeal mush. One of his specialties was vinegar pie. Not a bad substitute if you had no lemons, and that was the exact number we had.

Poets' Names and Poets' Faces

At a summer station on the Fillmore National Forest in central Utah, the work of the privy artist took on a new character. The predecessor of the WPA model was one of those three holers, with a large seat for Papa, a medium seat for Mama, and, at a lower level, a wee small seat for Junior. On the back wall and above the three seats were these inscriptions—"$4.00 Per Diem"—"$2.00 Per Diem"—"$1.00 Per Diem." In those days, per diem was paid instead of actual travel expenses. District (Regional) Officers, $4.00 per day; Supervisor and Assistants, $2.00; and Rangers, $1.00. This was a fair arrangement since visiting officers had hotel stops and hotel meals. The field men had many of their meals at sheep or cow camps or guard stations, or cooked their own groceries and used their own bedrolls. The Supervisor and Assistants were half and half. Each meal counted as one-fourth of a day and lodging was one-fourth of a day. Motels at fifteen dollars a night hadn't been invented then. If a ranger forgot to take a lunch with him, he could collect twenty-five cents per diem just the same. No per diem was allowed unless one was away from home overnight.

A Bachelor's Predicament

In the summer of 1921 Dave Shoemaker, of the Ogden Office of Grazing, and Mrs. Shoemaker crossed trails with me on Orange Olson's District on the Manti Forest. The four of us stayed overnight at Orange's summer station and he made the hot cakes. After breakfast Mrs. Shoemaker volunteered to do the dishes and was rummaging in the drawers. Then she said, "Orange, where do you keep your dishtowels?" With a perfectly straight face, Orange reached behind the cookstove and pulled down a disreputable, grimy old towel and said, "This is the only clean towel in the place."

Orange was promoted to Assistant Supervisor the following year and then served as Forest Supervisor for eight years. Following that he was in charge of wildlife work in the Intermountain Region for fourteen years. He met death in a plane crash in 1945 while making an aerial study of the Grey's River elk herd on the Bridger National Forest, Wyoming.

His friends published his writings in a valuable and interesting book entitled *Elk Below*,* as a memorial to him and his contribution to the cause of wildlife

New Overcoat

Arthur McCain, in charge of "Operation," and "Bish" Gery, in charge of "Lands" in the Intermountain Region,

---

* Publishers: Stevens & Wallis, Inc.

met in the hall upon Gery's return from an inspection trip to the forests. Said McCain, "Oh, I see you have a new overcoat. I expect you put it in your expense account."

"Well, if I did, I bet you'll never find it," answered Gery.

Bean Bags!

In Missoula, Montana, O. C. Bradeen built up "Central Purchase," for the Northern Rocky Mountain Region, into big business. It started as a supply base for fire fighting but grew to include almost all Forest Service purchases in the region as well as purchases for other agencies. It was an efficient and well-managed organization—but workers were sometimes human.

Ed Mackay, ranger on the Lolo (East Selway) Forest requisitioned sixteen ten-pound sacks of navy beans along with a big list of groceries for trail crews, lookouts, and district workers. The road down Crooked Fork ended seven or eight miles from the Powell Ranger Station and everything had to be mule-packed the rest of the way. One day one of the packers unloaded eight mule loads of white beans. Ed looked at the 1,600 pounds of beans and exploded, "What the hell do they think I'm going to do, feed the whole U.S. Navy?"

The beans were received at the warehouse in hundred-pound bags and put up for distribution to the Forests in ten-pound sacks.

Orange Marmalade

Bradeen got a bargain in orange marmalade in 1927 and stocked up to the gunnels with it. The forests send requisitions by ranger districts. A district order might be for peach jam and plum jam. The filled order showed "out of peach jam—out of plum jam—orange substituted." The next time the district would try for apple butter and apricot jam, but got the same treatment "Out of apple butter, out of apricot jam—orange substituted." You may love orange marmalade on your morning toast but there is a limit.

Recently I told John Rogers, supervisor of the Wallowa-Whitman, about the giant marmalade supply and he said, "That explains why we found a number of old cases of orange marmalade in a basement of the station at Deer Lodge, Montana, a few years ago."

Marmalade must have good lasting qualities. Twenty-five years—aged in the tin.

What's in a Name?

The folks in the Northern Rocky Mountain Experiment Station had a Numb Skull Club. Anyone who pulled a real boner was promptly nominated for membership and his reception was certain.

Kenneth Davis, who started as a fireman on the Lolo Forest and later became Dean of Forestry at the University of Montana, earned his membership this way. In a list of surplus government property at Washington, D.C., available free, he discovered a paper cutter. He decided to order it as it would save a lot of time trimming kodak pictures and other paper material.

He had almost forgotten about surplus property when he received a freight notice from the depot. With his travel sedan he went by to pick it up. To his horror he was shown an enormous hydraulic press cutter, weighing tons, and all by itself in a freight car. The freight from Washington, D.C., to Coeur d'Alene, Idaho, was $375.

Mistaken Identity

One July, as a delegate from Kootenai County, I attended the North Idaho Chamber of Commerce Convention at Orofino, Idaho. The State Mental Hospital is located there and the superintendent of the institution was on the program. At the conclusion of his talk he invited us to visit

Courtesy U.S. Forest Service

IT TOOK EIGHT PACK ANIMALS TO DELIVER THE NAVY BEAN SHIPMENT TO POWELL RANGER STATION, IDAHO

See page 319

the hospital. The chairman thanked him and urged us to accept, adding that he was sure the doctor would guard against our encountering any sordid cases. The doctor assured him there were no sordid cases.

Next morning, with two others, I called at the superintendent's office. He took us on a guided tour, through the dining room, kitchen, bedrooms, recreation areas, infirmary and headed down a long hall. We could hear a terrible commotion and, as we moved along, it became louder and more raucous. We caught each other's eyes, thinking the doctor had been too optimistic. It was becoming painful.

At the end of the hall the Doctor unlocked the door and unhesitatingly motioned us inside. We went in doubtfully.

It was the men's sitting room and forty men were quietly seated listening to the radio. It was bringing in the Democratic National Convention at Chicago.

Minding His Own Business

Retired Ranger Ed Mackay, Lolo Forest, tells this.

During the early 1920's we had two and sometimes three ten-man trail crews. In one of these camps we had an old cook, Chris Boding. Chris always used the front part of an old pair of bib overalls for an apron and seldom washed it. For convenience in seasoning food he would carry salt in one pocket and pepper in the other. At that time we began getting canned jam through Central Purchase at Missoula. Our first supply was fig jam. Old Chris opened a couple of cans and set them on the dining table we had under a 16' by 20' fly. Chris didn't deem it necessary to cover the cans to keep the yellow jackets out and the jam, being sticky, those that got in stayed. We had a gruff old lumberjack in the crew, Bruce Hunter. One evening at supper Bruce took a big helping of jam, and by that time it was about half yellow jackets. They were the same color as the jam and not too noticeable. A young college student working on the district for the summer sat next to him. Said the young man, "Bruce, you are eating yellow jackets."

Bruce never looked up but said, "That's their lookout, not mine," and never stopped eating.

Gas Troubles!

Frank Bishop was foreman of a road job on the Coeur d'Alene Forest one summer. The crew occupied a temporary tent camp. The gasoline was in fifty-gallon barrels. Gas kept disappearing. Without saying anything to anyone, Frank half filled an empty gas barrel with water and shoved into it the intake pipe of the small hand pump. The men in the crew took lunches and were gone all day. When they came in at 6:00 P.M. there was no supper ready and no cook in camp. The cook had his private car and evidently had been going out for drives. This time he must have coasted quite a distance because the gas in the line took him several miles down the mountain and when the water reached the carburetor he had a long walk.

Howard Phelps, from Operation in the Portland office, was on an inspection tour on the Whitman. He and I made a big swing through Sumpter, Granite, Chicken Hill, and Anthony Lakes with numerous side trips, and the gas in the little Hudson sedan was low. I was afraid we couldn't make it to a service station. On the way to the valley from Anthony Lakes we passed Bill Clark and his patrol grader and road maintenance crew. Farther on we saw his parked service truck with a gas barrel and hand pump, so we pulled over and stopped. We filled the tank, relieved.

Our relief was short-lived for the Hudson began to sputter and the motor died. It dawned on me that the patrol grader was the only government vehicle on the forest using diesel oil. A four-mile walk to the nearest phone at Christensen's Shadylane Dairy brought Ranger Kenworthy from North Powder to our rescue.

Shock!

In the summer of 1922, Supervisor Scribner, of the Salmon National Forest, and I were on a pack trip on the forest. One day we were headed down the trail on the north side of the Salmon River—the "River of No Return." We were headed for the Pope place, the last habitation down the river from Shoup. Scrib told me about surveyors for the railroad going down the river scouting the possibilities for a railroad location. When the squatter's wife was told of it, she directed her husband to start that afternoon cutting poles to make a fence between the house and railroad. She said, "I'm not going to have my kids run over by any train." No whistles have been heard to date.

We were laughing about it as we came into an opening and could see the cabin a third of a mile away. Shortly we saw a man walking hurriedly toward us with an old floppy hat in his hand. When still some twenty feet away he called excitedly, "I got word from Salmon that your wife has died."

Scrib exclaimed, "Whose wife?"

He said, "This man's," pointing to me.

Somehow I rode the three hundred yards to the cabin, slid off my horse and immediately slumped to a seat on a block. I've never been in real shock before or since. I couldn't see, I was chilly, I couldn't think, I couldn't feel. I have a dim recollection that Pope brought me a cup of sour wine that I gulped down. Time passed. What could I do? I was a two-day ride from a road, and two long days from there to Ogden, Utah, where I lived. I just sat and shivered, even though it was eighty in the shade.

After a couple of hours that seemed like that many days, Scribner was able to run down the report over the Forest Service tree line to Salmon City and he learned that it was my wife's father who had passed away instead of my bride of three months.

CHAPTER SIXTEEN

MEN WHO BUILT THE FRAMEWORK

Leon F. Kneipp, Regional Forester

I WAS UNDER LEON KNEIPP'S SUPERVISION SEVEN YEARS, the entire time he was District Forester (Regional Forester) of the Intermountain Region, Ogden, Utah.

He had a wonderful command of the English language and was a fluent speaker. Yet he told me one time that his legs always shook like aspen leaves whenever he talked in public. That was real encouragement to me as I suffered from platform ague when exposed to an audience.

He only once even gently laid my ears down. At an allotment conference I proposed the purchase of packsaddles for ranger use as they often had to furnish their own equipment in packing government telephone wire, barbed wire, equipment, and camp supplies. He passed over the subject and later I brought it up again. He made it plain that he heard me the first time.

As a lesson of this sort Jack tells of attending a country auction. The skilled auctioneer was trying to start a pen of a dozen geese at a ten-dollar minimum. A man persistently tried to start it at "one dollar." The auctioneer couldn't hear him and a helpful lady kept saying, "Here's a bid over here, Mr. Thorp." Finally the auctioneer stopped, turned to her in a wonderfully friendly way, and said, "How are all your folks, Mrs. Underhill?" The crowd laughed and someone bid "Ten dollars."

All the old-timers in and out of Ogden knew District

Forester Kneipp's handwriting was beautifully back-slanted. His signature was a work of art, what with several extra flourishes.

A stranger might have difficulty in making out the initials, "L. F." One correspondent who found himself in that predicament worked out this solution. He cut out the signature, pasted it on the envelope, and mailed it to Ogden, Utah. Alvin Johnson, the mail clerk, spotted it and showed it to Clyde Gwinn, Chief of Maintenance. Gwinn cautiously showed it to me.

Major Evan W. Kelley, Regional Forester

I first met Evan Kelley in 1919—that's forty-seven years ago. That was in Ogden, Utah, and he was compiling a For-

Courtesy Ed Mackay

REGIONAL FORESTER EVAN KELLY ACCEPTS RETIREMENT GIFT OF PACK MULE AND OUTFIT FROM MISSOULA KIWANIS CLUB

The Governor's Ballroom, second floor of the Florence Hotel. Ed Mackay (*left*) and Bill Bell (*right*) took the mule up in the elevator.

est Service Trail Handbook. Then I saw him in action at the Mather Field Fire Conference, Sacramento, in 1920. He was Fire Chief in the Washington office at that time.

In 1922 he and I made an extended pack trip on the Boise National Forest, Idaho. It takes a camping trip right out in the forest to get acquainted with a person. I found him to be a good horseman, a good forester, and a good cook. One night he produced a dish I'd never tasted before and no one could ever guess: Tomato dumplings cooked in a Dutch oven. Good, too.

When he came to Missoula, Montana, as Regional Forester, I was there as Supervisor of the Lolo Forest. For the next fifteen years he was my immediate boss. During that time I saw him in action in the battle of the CCC's. The white pine blister rust campaign was started and grew to big proportions. The fires of 1929 and other years were fought. The Remount Depot was set up. Hundreds of miles of bulldozer roads were pushed out to help the fire job.

Evan was not a forestry college graduate but he was as good a forester as they come. He was a keen observer—he didn't miss much. He was a voracious reader and remembered what he read. He developed and grew and broadened himself more than almost anyone I know. He was a man of action.

Lyle F. Watts, Chief Forester

My association with this man has been longer and perhaps closer than with any other forester. He and I were the only two Forest Assistants appointed in Region No. 4 (Ogden) during the year 1913. He came from Ames, I from the University of Minnesota. We spent a couple of winters together in the Ogden "Bull Pen" on lower Twenty-fourth Street.

One evening over a "Greenriver" at the Falstaff we decided to stag it to a party in the Congregational Church. After a hilarious evening playing drop the handkerchief and other parlor games, he walked Nell home and I walked her friend Blanch to her home. Lyle's case took, Nellie Bowman later becoming Mrs. Watts. Mine misfired. However, I did marry Ruth, another frequenter of that musty old church basement.

While he was Supervisor of the Idaho National Forest, I was working out of Ogden.

While I was Supervisor of the Lolo National Forest he came to Missoula as Director of the Northern Rocky Mountain Experiment Station. One of his projects as Director was the establishment of the Deception Creek Experimental Forest Area, on the Coeur d'Alene Forest, while I was Supervisor there.

From Missoula he became Regional Forester of the Pacific Northwest Region at Portland, Oregon. In 1940 I moved to the Whitman under his supervision.

Courtesy Mrs. Lyle Watts

TIMBER CRUISERS

Lyle Watts and J. W. Stokes on the Targhee National Forest in Idaho, 1914. They missed the biweekly stage.

He was chosen Chief Forester, Washington, D.C., in 1943, and held the job until he retired in 1952.

If I could use only five adjectives to describe him, they would be "sharp," "farseeing," "stable," "forestry minded," and "friendly." Lyle made friends for the Service in all walks of life wherever he went.

Colonel W. B. Greeley, Chief Forester

Colonel Greeley was our third Chief Forester and held that important post for eight years. He then resigned and continued in forestry work for private owners in the Pacific Northwest. I didn't know him well, though I had several personal contacts with him. He was one of the very few right up near the top of my list.

Here is a quotation from Stewart Holbrook's article in *American Forests* magazine of March, 1958. It is titled "Greeley Went West."

"And now, in the spring of 1928, at the age of forty-nine, with two successful careers behind him, William Buckhout Greeley had come West to start a new career. Officially he was secretary-manager of the West Coast Lumberman's Association. Actually he became and remained to his death the outstanding leader of the lumber industry of the United States. . . .

"Yet no other man did more, or perhaps as much, to warrant the honor which Yale University plans for its distinguished son. The William B. Greeley Memorial Laboratory will serve to remind generations to come that surely, in those days, there *did* walk giants in the land of Goth."

Gifford Pinchot—Henry Solon Graves—Herbert Smith

One outstanding event of my career was the visit of these three men to the Coeur d'Alene Forest in 1938. As host supervisor I was privileged to ride in the Governor's personal car with the three and a Negro chauffeur. Regional Forester Kelley and regional officers followed in a second car.

Gifford Pinchot was the first Chief Forester in the United States. He organized the Forest Service in 1905 and set its course, with the confidence and full backing of President Theodore Roosevelt. And he had since been a successful governor of Pennsylvania.

Henry S. Graves was first the Dean of Forestry at Yale University. He wrote a book, *Silviculture*, that became an outstanding text, used in every forestry school. As the second Chief Forester he served for ten years, carrying on the tradition established by Pinchot, building on his foundation.

Herbert Smith was an ex-newspaperman who was recruited by Gifford Pinchot to arouse the American public to the need for care in the prevention of forest fires and for a strong forest organization to manage the Federal forest lands. Smith headed up the vast publicity campaign beginning in 1905 in support of conservation.

The trio were making a tour of some of their old stamping grounds, to relive their earlier experiences and observe the fruits of their efforts. All three were out of harness and relaxed. It was both a reunion and a homecoming for them.

We showed them some of our dozer-built fire roads and some of our higher standard roads. They were amazed at the increased accessibility. We showed them some of the clear-cut and controlled burn areas of the overmature white-pine, hemlock, white-fir type, some plantations, and recent and earlier selective cuttings of white pine. They saw the white pine experimental work at Deception Creek near the Honeysuckle Ranger Station. They reviewed our fire organization and records.

We stopped for the night at the Deception Creek Station and everyone enjoyed the clean beds provided for visitors— that is all but two. Believe it or not, Gifford Pinchot brought out a sleeping bag from his car and walked a couple of hundred yards up a trail, found a spot that looked good, and

Courtesy Charles Simpson, Baker, Ore.

GIFFORD PINCHOT, HENRY S. GRAVES, AND HERBERT SMITH VISIT THE COEUR D'ALENE NATIONAL FOREST AND LOCAL AND REGIONAL FORESTERS, 1938

The roster, starting at the left, is: Howard Drake, Logging Engineer; Gifford Pinchot, first Chief Forester; Evan Kelly, Regional Forester; Herbert Smith, ex-Assistant Chief Forester; C. K. McHarg, Assistant Regional Forester; Kenneth Davis, Research Experiment Station; Hans ———, German Forester; Steve Wyckoff, Director, Experiment Station; Charles D. Simpson, Forest Supervisor, Coeur d'Alene National Forest; William G. Guernsey, Assistant Forest Supervisor, Coeur d'Alene National Forest; James Ryan, Supervisor, Kaniksu National Forest; Elers Koch, Assistant Regional Forester; William Larson, Ranger, Magee Ranger District; George Haines, Supervisor's Staff; Driver of the Pinchot car; Harry S. Graves, second Chief Forester and Yale Dean; Virgil Moody, District Ranger, Coeur d'Alene District.

bedded down on the ground for the night. Regional Forester Kelley kept him company.

Next morning the cook served a fine breakfast. Among other things on the table was a bowl of stewed prunes. The Governor helped himself liberally and praised the cook. He added that his initials stood for good prunes. As we were leaving the cook presented the Governor with a three-pound sack of dried prunes which he accepted gladly and stowed in his car.

This will always stand out as a red-letter day. The Governor was still a most interesting personality. Contact with him made it easier to understand how he instilled so much into the embryo organization and why it could be projected forward to succeeding generations of foresters. Few persons in America have put a lasting spirit into an entire organization the way Gifford Pinchot and his early associates did. Enthusiasm of their rare quality rubs off on everyone they touch. Just enthusiasm alone isn't worth much. I have seen it expended on dubious projects. But with a happy combination of enthusiasm, personality, and a worthwhile cause, there is no limit to accomplishment.

CHAPTER SEVENTEEN

TREES AND TOWNS

E. R. Jackman

Trees and mountains go together, so most of the national forests are above the four-thousand-foot line. Ponderosa pine needs at least sixteen inches of annual precipitation and other kinds of trees need twenty inches or more. Junipers defy all rules and grow on favorable sites right down to ten inches of rain out in the desert. Most Western valleys have less than twenty inches of moisture so they have few forests, but nearly all of the mountains are tree-covered.

Through the ages since the earth's crust became solid and reasonably stable, and in the process wrinkled, folded, and faulted to form mountains, erosion has had its way. Streams and heaving ice brought soil from the peaks to the valleys. Lakes formed there that caught and held the soil. These valleys, often former lake beds, are fertile and are farmed. Much of the West is a succession of large or small valleys with tree-covered mountains or spurs of mountains between. The forests were divided into the various national forests, usually separated from each other by natural boundaries.

Each valley has a town, often the county seat, and each forest is administered from a headquarters in one of these towns. Therefore, geography, climate, and terrain all conspire to force United States National forest headquarters into towns usually between 2,000 and 10,000 in population. You don't find many individual forest headquarters in cities such as Los Angeles, Portland, or Denver. They will be in towns such as Bend and Baker in Oregon, Grangeville and Salmon in Idaho, Hamilton and Kalispell in Montana, and Fort Collins and Pueblo in Colorado. All of the national forests in turn are grouped into regions and regional headquarters may be in cities.

I think that a town of maybe five thousand population that is the headquarters of a national forest and is in a valley surrounded by tree-covered mountains, is, essentially, a better place to live than a town of equal size in a non-forested state.

Population of a valley, a state, or a region, depends directly upon the primary income. In the West, that income is usually from farming and harvesting of timber. The town that has both is in a better position than the town with just one. The money poured into the town from farms and forests supports the lawyers, doctors, grocery stores, hairdressers,

Photo by Gambs Studio, Baker, Ore.
INTERRELATION BETWEEN THE ECONOMY OF THE COMMUNITY,
THE FARM, AND THE FOREST
This farm, in Baker Valley, Oregon, draws irrigation water from the mountains. Posts, and the old split rail fence grew up there, too. Fat cattle develop on the spring and fall range grass and on the summer growth on forest lands.

service stations, and all of the other dozens of occupations. The valley that produces yearly $20,000,000 from these two primary sources will normally have twice the population of the one with $10,000,000.

Most of the income from the national forests goes directly into the bloodstream of the local economy. The labor in both woods and mills is local labor. Local truckers do the hauling and truck maintenance.

In years when weather or prices, or both, conspire to reduce the flow of money from farms to town, it is likely that the forest income will stay high, so local business firms still live. Occasionally forest products are in the economic doldrums, but maybe cattle are high in price. This gives stability. For example, Baker, Oregon, has *never* had a bank

Photo by Photo-Art Commercial Studios, Portland, Ore.
THE HIGH RANGE, WITH AMPLE WATER, FORAGE, AND SHADE,
PUTS THE "BLOOM" ON THE LAMBS
Fat lambs help the economy of any community

failure. In the depression two banks merged, but none failed. Few business firms have failed there. It is the head-quarters of the large and productive Whitman National Forest (Wallowa-Whitman since 1952). Nearly every rancher is a cattleman.

The ranch activities are metered out by the seasons. Most of the ranchers graze their cattle during the summer on the national forest, following the feed supply. The ranches supply feed for spring, fall and winter. About May 30 stock are turned onto the forest where they stay until about October 1. In the fall, usually just before deer-hunting season, the ranchers get their stock out of the woods and back to the ranch again.

Every rancher has his feed needs carefully calculated. If he has more summer feed than he needs, he will buy some young stock in the spring, feed them his surplus, and sell in the fall. If he has more winter feed than he needs, he will buy feeders in the fall and sell them four months later. The weather is his unforeseeable factor. It can double his carrying capacity or halve it. It is his partner, and may at times be his enemy. It can make or break him.

For such a man the feed on the forest is a great stabilizer. Rains in the mountains are more dependable. If short of feed at home in the spring, he can provide concentrates, or rent pasture, or corral feed. But when he turns onto the forest, he then has a breathing period of four or five months. In drought years he can sell stock, buy hay, buy concentrates, or ship part of his cattle to some other county.

In any case his operations are essentially a means of turn-ing feed into money. The more feed, the more stock, and the more money. The "Cattle King" and "Cattle Baron" talk is mostly myth. In county after county *all* the cattle are owned by individual ranchers on family farms. The parts of America formerly owned by Spain have some king-sized ranches. They are hard to find in the homestead states.

As mentioned, there is some elasticity in operating a cattle ranch, but a cow eats about the same amount every day of the year if she can get it. So a farmer deprived of usual feed in any series of months is in a bad fix. He doesn't make any money that year and if he has a series of such years he goes

Courtesy E. R. Jackman, Corvallis, Ore.
RIPE PONDEROSA PINES CONTRIBUTE THEIR SHARE TO
COMMUNITY PROSPERITY
Wages, machinery, gasoline, tractors, trucks, mills, railroads—all are needed in the operations which take place between the time when the tree is cut and the finished lumber and manufactured wood products are ready for market.

broke, or sells out to a California developer or a Texas oilman.

But back to the towns. These small county-seat towns have been stable for about one hundred years. They are fine places to live. The presence of the national forest in itself helps them, because in any day of the year the townspeople can look up to the mountains, see what the weather is up there, and, if suitable, off they can go for stimulating, healthful, and wonderful fun in the woods or along the streams and lakes.

Chapter 9 in this book treats of what the forests offer, and I won't go into it here, except to remake the point that a town so located has a great advantage as a place to live.

The presence of the United States Forest Service organization is another help. These forests differ, but usually have around ten regular employees. These men are assets to any town. Every town has myriads of public-service jobs that call for community or group activity. Just to list a few:

Scouting and 4-H work—training, hiking, camping, tours, summer camps
Pageants and fairs
Church and civic club activities
Maintenance of picnic grounds and local fair or rodeo grounds
Disaster help—fires, floods, winds, storms
Lifesaving activities and training
Rescue parties for airplane crashes, lost persons
Clean-up days
Tourist information and greeting of visitors
County tours or large meetings, such as conventions

There are hundreds of such things in every one of these towns. Taking care of non-paid jobs that someone must do is a real tax on one's time. The community that does it well creates a good image, the town that does it poorly or not at all is a poor place to live.

The point here is that these ten or more United States

Forest Service employees tend to absorb many of these jobs. They are well educated, of uniformly high character, and many have special skills. This man may be an expert photographer who can train young persons and can give interesting illustrated talks to garden clubs. Another is a fine horseman, an efficient packer, and he can take mounted groups into the high mountains. A third is as good as you find in training lifesaving crews. One man is a fine singer and can organize quartets.

The Forest Service personnel changes from year to year as retirements, promotions, and special service needs occur. So every forest headquarters town, through the years, has a procession of these public-spirited men. The community life is richer and more enjoyable because of them.

Charlie Simpson's life illustrates how a Forest employee fits into a town. Aside from serving on hundreds of com-

Photo by Parker's Studio, Libby. Mont.
THIS SEVENTEEN THOUSAND FOOT LOAD OF LOGS
MEANS MONEY IN THE BANK

mittees and special-duty groups, some of his official positions have been:

President—Intermountain Section, Society of American Foresters, Ogden, Utah

President—Northern Rocky Mountain Section, Society of American Foresters, Missoula, Montana

President—Rotary Club, Coeur d'Alene, Idaho

President—Chamber of Commerce, Coeur d'Alene, Idaho

Vice-President—North Idaho Chamber of Commerce

Director on Board of Trustees that founded North Idaho Junior College at Coeur d'Alene, Idaho

President—Kiwanis Club, Baker, Oregon

Trustee and Elder—Presbyterian Church, Baker, Oregon

General Chairman of Building Committee that built the Presbyterian Church in Baker

Courtesy E. R. Jackman, Corvallis, Ore.

A TWO-WAY BENEFIT

Youngsters at 4-H camp enjoy the week-long experience of tent living, good food, clean air, crystal water, and cool nights. They are also benefited by instruction in safety in the woods, provided by a qualified public-spirited forest ranger. Fremont National Forest, Cottonwood Meadow Recreation Area west of Lakeview, Oregon.

Chairman—Eagle District, Boy Scouts of America
President—Blue Mountain Council, Boy Scouts of America
Recipient—Silver Beaver Award for distinguished service
   to boyhood
Lieut.-Governor, Division 17, Pacific Northwest District
   Kiwanis International

But the important thing is the forest itself. It doesn't take
a John Muir or a Henry Thoreau to enjoy it. Every man in
Baker gets something from the mountains. In less fortunate
communities, a mountain trip is something to look forward
to for years. In such a town, a man usually has everything
ready, and from the time he conceives the idea to the time
he is in the woods with his family may be only thirty minutes.
   It doesn't matter much what his forest enjoyment is. The

Courtesy U.S. Forest Service
FROM THE HAVES TO THE HAVE-NOTS
Christmas trees from the Kootenai National Forest, cut in the far northwestern corner of
Montana, are on their way to the big treeless valley of the Snake River in southern
Idaho. How fortunate are the families who, in an hour or two, can go out on the
forest and cut their own tree and Christmas greenery.

science-minded man may be a botanist, geologist, ecologist. The nature lover can study trees, animals, birds, or wild flowers. The busy businessman may relax in the beautiful silence. The office worker may inhale the pure fragrance of the green-growing things. The forest has something for every person in America—something that will enrich his life and grow with each visit.

The well-to-do of Illinois or Ohio may have a retreat in Colorado or Canada. Every auto mechanic or store salesman in one of our towns may have what only the rich can enjoy elsewhere. I have lived my whole life with a low income of an average college professor. Even in the depression we could be in some lovely vacation spot in a matter of minutes —a different spot each weekend for our entire life if we wished.

Families brought up that way have no delinquency problems. There is no essential difference between rich and poor if both like the woods and streams. The fine mountain trout, lurking in the deep pool behind the big rock doesn't care how much money you have when he strikes your fly as it swirls in the eddy. The antlered deer or elk is as likely to be brought in by the janitor as by the bank president. Nature is a great asset to a democracy. Any man can be king as he watches the shadows creep up the mountainside from the dark canyon, or as he sees a doe daintily pick her silent way through fern and wild flower to the still pool.

Our western forests help the towns, whether one is business oriented or is poetically minded. The fortunate man is both and I think in our forest-girt towns one finds a higher percentage of such persons.

CHAPTER EIGHTEEN

TREES AND HUMANS

E. R. JACKMAN

MOST OF US DO NOT LEARN MUCH FROM LIFE. AS FISH take on protective coloring as they become older, we learn how to keep from getting hurt, how to live with others. We learn not to laugh at the wrong time, when to be silent. Such things are good for one's business. But they do not make us bigger; they just make us more careful.

Life in a city, or even a town, isn't too good for us. A jangling phone, a blasting horn, the blasphemy of a jukebox, the hundred interruptions and irritations of the daily business life—these things do not build character or provide peace of mind.

We have created a strange civilization. All of the backward nations envy us, but often the old Basque shepherd on his age-old hills seems to have a true wisdom hard to find on Broadway. Wisdom is more likely to come in solitude. We can find it in the forests, the desert, and, in some places, by the sea. We can rest tortured nerves and cure doubts and maybe make ourselves more worthwhile as companions for ourselves—and others.

Thoreau's name shines more brightly with time. He learned truth in a far less complicated age, over a hundred years ago. He said:

"I went to the woods because I wished to live deliberately, to front only the essential facts of life, and see if I could not

learn what it had to teach, and not, when I came to die, dis-
cover I had not lived. . . ."

And, on another page of *Walden* he said,

"I never found the companion that was so companionable
as solitude."

Charles Darwin, in his journal during a sea voyage, re-
flected:

"Among the scenes which are deeply impressed on my
mind, none exceeds in sublimity the primeval forests unde-
faced by the hand of man. No one can stand in these solitudes unmoved. . . ."

Of course poets have found beauty and solace in the
forest. Bliss Carman was a Canadian poet and the woods
brought reverent moods to him.

> "There is virtue in the open; there is healing
> out of doors;
> The great physician makes his rounds along the
> forest floors."

and

> "I took a day to search for God
> And found Him not. But as I trod
> By rocky ledge, through woods untamed
> Just where one scarlet lily flamed,
> I saw His footprint in the sod."

From Shakespeare to Joyce Kilmer the great poets have
expressed these thoughts. That is the special virtue of poets;
they say the things that we only feel. Most of us feel the
same things, but in a vague, confused way, almost ashamed
to catch ourselves with a poetical thought.

If we have half formed feelings of reverence in the solitude of the forests, we are in good company. The great religious leaders of all time developed their creeds in solitude. Buddha, reared in luxury, was troubled by doubts. He retired for meditation and fashioned his religion only after years of solitude. He laid no claim to being a god. He was only a man teaching tolerance and calmness.

Confucius also sought solitude and emerged after many years to teach love and tolerance. He, too, laid no claim to divinity, but his teachings ruled China for two thousand years. While other religions led to strife and wars all over the known world, the "inner calm" taught by Confucius kept the Chinese a peaceful people.

Mohammed came a thousand years after Confucius, but he, too, developed his beliefs in solitude and wrote the Koran, though he started life as an illiterate poor boy.

Jesus often repaired to solitude to pray, although he did not stay for long periods.

Courtesy U.S. Forest Service

PONDEROSA OR YELLOW PINE

In the Fremont National Forest of Oregon. Cool, fragrant, restful for the soul and body

The quiet, calm, and patient trees can teach us things if we let them. We cannot learn those things in traffic jams, in crowded elevators, or in the midst of whirring machinery. A forest is a good place to achieve balance, to put things in our minds in their proper importance. All about us is ruled by reason. This tree cannot grow well for the soil is shallow; this tree is weak for it is in the shade of another; in this wet spot is a little spring and in the saturated soil only aquatic plants can develop. There is a reason for everything one can see and this order corrects the disorders that the hurries and jangles of civilization have crowded into our minds.

Effect of Woods

Poets do much with forests and the trees in them, but the folks who actually work in the forests do not normally maunder around repeating with Joyce Kilmer that "only God can make a tree." It would look pretty silly in an official report—sort of on the mawkish side.

I remember a guide in Glacier National Park. He was in charge of a mounted party. One member was an Eastern lady who had never ridden before. She had consumed far too many chocolate eclairs in her medium long life and could not get on her horse without much pushing and hoisting by the guide. All day she had insisted upon getting off to marvel at the frequent views, with a struggle by the guide almost beyond his weight-lifting powers.

Late in the afternoon the party rode past a tree rather large for that country. The lady wanted to get off, and the guide helped her down. She walked over to the tree, brought her waistline to attention, raised her hand, and said solemnly, "Oh tree! I salute thee!" The guide, a local boy, didn't say anything, aloud, but his lips moved.

A forester is apparently changed by his life, for the living majesty of the woods does something to him. I have known many men in many of the burgeoning government services and, according to my standards, the personnel of the Forest

Service far surpasses that of any other Federal agency. Of the
hundreds I have known, none has been mean.

Chatterboxes among foresters are rare; disturbing egotists
are seldom found among those actually out in the forests;
and ignoramuses are nonexistent. The only criticism that I
can make of any I have known is that occasionally one is
so dedicated to his work that other things lack importance.
This throws his views out of perspective and sometimes irri-
tates those in other professions.

I do not know whether the very nature of the work at-
tracts good men, or whether the forests gradually erase the
bad traits and build the good ones. This is certainly not the
case with many city jobs. Some tend to develop clock watching, jealousy, selfishness, impatience, intolerance, and all
of the things that degrade humankind. Just read the letters

Courtesy Oregon Extension Service
THIS 4-H CLUB CAMP STAFF HOLDS A COUNCIL MEETING IN THE SHADE
OF THE PINES DURING BAREFOOT HOUR AT COTTONWOOD SUMMER
CAMP, FREMONT NATIONAL FOREST
Staff members are, *left to right*, Donna Holloway, Harold Kerr, Mrs. Virginia Ray, Lee
Hansen (county agent), Mrs. Ellen Hawk, Mrs. Mary Padget, Earlene LaBranch, Alma
Crowl.

to the editor in any metropolitan paper. I have stood upon
busy corners in Chicago and Pittsburgh and scanned the faces
of the hundreds of passing pedestrians. They show all kinds
of traits, but serenity, tolerance, and a quiet competence
seldom appear.

Different forests have completely different characteristics
and make their own impressions. I think that the most callous person feels some awe in the overpowering "cathedral
aisles" of the redwoods, "lovely, dark and deep," as Robert
Frost, the poet, said. A virgin stand of Douglas fir on the
coast where the sunlight has not shone for two hundred
years causes almost the same feeling as a dark, cool church.
It is outside the experience of many and a stranger to it may
feel a vague disquiet, much as a Plains Indian might feel in
an unlighted house of worship. But when a wanderer raised
in the firs, returns after a long absence he fairly bathes in
the cool depths, feels at home and at peace.

Some students of population and climate say that the
ponderosa pine and humanity have identical requirements:
that if the climate is too hot, too cold, too wet, or too windy
for our yellow pines, those same things retard the best development of human beings. The open, parklike ponderosas
breed no feelings of oppression or claustrophobia. The sun
shines in them, and wild flowers grow. Deer are present,
many birds are happy all about, and children can run and
play.

We don't have much birch in Oregon. Canada, Alaska,
the Lakes States, and the Mountain States have real birch
forests. A birch in the spring is one of the most graceful and
beautiful of all trees. It is poetry of both motion and color
and is nature's description of spring, newborn and joyous.
Dancing maidens, young and beautiful, might well wear the
colors of the birch, white and a peculiar light vibrant green.

Oregon has millions of acres of lodgepole pine. Its effect
is completely different. It tends to grow where it has cold
wet feet part of the year, and neither trees nor humans really
enjoy that condition. So a forest of lodgepole, as the Latin

name signifies (*Pinus contortus*), has many imperfect trees; they die young, often resulting in such a tangle of down trees that walking is slow and riding a horse may be impossible. The bark is rather scaly and unattractive. I wouldn't enjoy living in the middle of such a forest, or at least ponderosa pine makes a better place in which to live.

Most of the other kinds of trees in the West do not grow in pure stands except in highly specialized conditions. In Oregon we have the friendly quaking aspen groves at high elevations, such as our Steens Mountain. The Port Orford cedar, tall and straight, lives in its limited chosen coastal hills. And of course the juniper grows on every elevated place in the big sagebrush country of the high desert. It is hardly thought of as a forest because most of the trees are unfit for lumber, for the wood is full of long gaps or hollows—fine for wild bees, but impossible for ordinary boards.

Most of the various cedars, the other several pines, and the firs, except Douglas, are rarely in pure stands—the yew never is. The white pine of North Idaho is sometimes in large stands, mostly pure. There is a good deal of western larch, but it grows on the north slopes mixed with other species. It is a curious conifer in that it sheds its needles in the fall and passes the winter naked as a poplar.

A story was once told to me about the larch; tamarack to most residents. I cannot vouch for it, but it sounds as though it might be true. A young forest employee came to Idaho, his first job out of school. He was sent far up into a North Idaho forest in the mountains. He was the nearest Heaven he ever expected to be; deer and elk on the meadows in early morn; fish in the waters; beaver in the creeks. He had not known there was such a place. Everything interested him and he wrote long reports to his superiors.

But when fall frosts killed the beautiful light green needles of the tamaracks, dotted all over the hillsides among the firs and pines, he sent a hurry-up call to the Forest Supervisor: "Send up an entomologist and a plant pathologist;

the western larch is all dying." The story was said to have followed him wherever he went.

Humans love the forests, partly because in this western mountainous country, the lowlands tend to be dry and hot in the summer, so forests can't grow except on the hills at a higher elevation. The elevation makes for cooler air, and that in turn causes precipitation, snow in the winter, rain in summer, so the dry-land wheat farmer sees his crop hurting, scans the blue sky above him, and looks longingly at the forested mountains. It is often raining there, while dust devils chase each other across his summer fallow.

That is one reason why, in our country, every rancher and every town dweller tries to get a place "up on the mountain." It may be just a patch, or large enough to pasture cattle. But he and his family can repair to their alternate mountain home when hot weather drives them away from valley or plain.

Every community has its quota of persons who are different. Milton-Freewater, north of Pendleton, Oregon, has Stanley Caverhill, rancher and philosopher, who looks at the human scene with amused tolerance. He is frequently asked to speak at civic group meetings in the towns around.

At a meeting at Pendleton he said something like this:

"When our forebears came to this country and all the land was free, they didn't settle on the rich wheatland in the valley, they had only horses or oxen, and they homesteaded the timberland on the foothills and mountains. They wanted a place with wood and water, both prime necessities. They felled the trees and made their log houses and barns right there. Then they started a lifetime of hard work clearing off the rest of the trees so they would have farmland.

"In the meantime the railroads came, lumber could be shipped in, and the second wave of homesteaders took up the good land down here toward Pendleton. Towns grew up and everybody used lumber for houses. They thought the

Photo by Photo-Art Commercial Studios, Portland, Ore.
IN MOST PEOPLE, A LOVE OF THE OUT-OF-DOORS RUNS STRONG
A week or two in such surroundings as these can correct the damage to the nerves caused
by the blare of horns, the hysterical alarms of the news commentators, and the rush and
danger of traffic.

old log houses up in the hills were picturesque, but gosh, they wouldn't want to live there.

"So the original homesteaders had an inferiority complex and they worked their fingers to the bone, slaved, and denied their families luxuries so they could accumulate enough money to move to town and live in a frame home.

"You people of the Rotary Club are their descendants; they were your grandparents. And now all of you live in nice frame houses and you work your fingers to the bone, slave, and deny luxuries to your families so you can accumulate enough money to get a log house up in the hills."

It was true—most of the persons there either had such a place or were working toward it.

Photo by Oregon State Highway Commission
MOUNT HOOD
Bear grass and rhododendron in the foreground. Tired eyes and frayed nerves return to normal amid such scenes.

So the folks in all of these Western States where timber covers the mountains have a deep interest in the woods and they tell their friends in Iowa or Kansas, "You can live in a prairie state of you want to, but there isn't enough money to hire me to live in a country without forests and mountains."

Having done considerable camping, I've noticed an odd thing. If a man camps on a plateau or meadow that is not forested, but has one solitary tree on it, he will carry his gear or drive over ditches and rocks to camp under the tree. This isn't for shelter, for such trees are too scraggly for shelter. There is something inside a person that wants a tree for company. Actually that isn't a very good place. Most lone trees get struck by lightning sooner or later. And if the land has been pastured, the spot is pretty well spoiled, for all the cattle want a tree at their backs, too.

CHAPTER NINETEEN

MY FRIENDS, THE TREES

ALICE S. HUTCHENS*

TREES ARE PERSONALITIES. AS YOU ASSOCIATE WITH THEM you come to know them, and what's more, when you have become acquainted with one of them, the memory of it will stay your whole life through.

In Minnesota, when I was very young, I knew two trees well. One was a butternut that overhung a stone wall at the end of our driveway. In its shade on top of the wall I used to set out blue-green plates made from trumpet honeysuckle leaves and acorn cups that I kept in a crack in the wall. Sometimes an old Scotsman came there in his kilt and played his bagpipes under the tree. Grandmother would send me down with small coins for him. I used to wait there for Andrew, the hired man, to come in from work in the field. He would toss me up onto the back of big brown Rob and let me ride to the barn. In fall, I helped gather butternuts there for Mother to use in a special cake she made.

The other tree was a large sugar maple that grew in the yard. It had a scar on one side where sap ran down in spring. Butterflies, bees, and moths came to it and got grounded in the sticky sap. It was a source of wonder for a small girl. There are so many of the loveliest things that children aren't allowed to touch, but here I could feel the gorgeous satiny

---

* This is a letter from Alice S. Hutchens, of Kalispell, Montana. The previous chapter of this book is an attempt to relate the effect that trees have upon the lives of those who live close to them. This letter, from a sensitive person, tells that story far better than my words can do. E. R. JACKMAN.

wings of the butterflies and stroke the soft furry bodies of the great moths and the bumblebees. Orioles always nested in that tree. Once they built their nest on a low-hanging limb. I could see right into it. It had one of my hair ribbons woven into it, along with some scraps of bright yarn that Mother donated.

When I was five, we moved to California. During the three years that we lived there, we children had four special trees. One was a huge, wide-spreading umbrella tree, a nice comfortable tree for climbing. We spent hours clambering around through that tree. When I had learned to read, I had a favorite reading perch up as near the top of the tree as I could get.

There was a tree that Mother called a "saucer-peach." The fruit was pink and white and early to ripen. That tree grew behind our carriage shed. We could climb up there and simply stuff ourselves with peaches provided Mother didn't catch us. She had other plans for saucer peaches.

The other two trees were mostly yours, Russell, and I'm not so sure that they were truly friends of yours. Do you remember the black fig? Only one branch, a long horizontal one, had any fruit on it. We couldn't reach them from the ground but you got up there some way, stretched out along the limb, and just as you were reaching for the figs the limb broke. You and it came down together so hard that all your breath was knocked out and you had an awful nosebleed. Mother was so frightened that she never did get around to punishing any of us.

We had a tall pine of some exotic species, with perfect whorls of branches growing all along its trunk. You climbed it once, and then had to stay up there until Father got home and borrowed the longest ladder in the neighborhood to get you down. Those whorls weren't very close together, the ladder wouldn't reach, and the branches up where you were wouldn't hold Father. I only remember how awfully frightened we all were. They got you down some way, for there you are, right now.

There's another thing about that pine. The very tip-top of it was a mockingbird's favorite singing perch. During the Spanish-American War everybody on our street was whistling "There'll Be a Hot Time in the Old Town To-night!" The mockingbird learned to give a perfect rendition of the first two bars of the song.

When we came to Montana in 1900, the farm that Father bought had on it two of a row of five curious, deep potholes with water in them. They were in our pasture which was still in prairie sod. Fir trees and aspen grew around them, with thickets of hawthorn, wild rose, and serviceberry bushes. Naturally, since there were no other trees on that place, we children loved those trees but were forbidden to play alone there because high banks went straight down to the water, and then down into it deeper than we could see. We managed to get down, though, every time anyone was working nearby. Exploring around under the trees, we found many Indian artifacts—round stones with a groove around them; long pointed ones with the large end rounded; lance heads; arrowheads of several sizes; and stone knives. Once we found a nest under the rosebushes. It had a dozen or more white eggs in it almost as large as guinea eggs. Each one had a neat brown circle around the large end. Something hatched there. We found the empty shells in the nest, but I've never known what bird it was.

I think I know, now, what the potholes are. In the house yard was a large odd-looking stone so heavy we couldn't budge it. We finally attacked it with a sledgehammer and broke off a piece. It was crystalized iron, such as some meteorites are made of.

Last week I drove past that place. The fifth and shallowest of the five holes is near the road. There is plenty of water still, and many kinds of waterfowl were there, even two Canada geese. The trees around our own two potholes were a real magic forest to us.

In later years, at the homestead at Star Meadows, just across a little creek from the cabin, was a small grove of jack pines. They stood out in the center of the field we had cleared, and under them was a deep carpet of moss and forest duff. I brought little ladies' slippers from the "big woods" and planted them there. When we moved from the homestead, the whole ground under those trees was covered with the flowers—a beauty spot not to be forgotten.

To me, birches seem the most beautiful of all trees. In winter they make so delicate a tracery against the sky. In spring their catkins and tiny new leaves are like Japanese drawings. In summer the sun-dappled shade under them seems cooler than under other trees, and in fall they are clothed with a glory of bright leaves. The ground under them always has ferns and delicate wild flowers. In May morels can be gathered there.

Photo by Bob Bailey, Enterprise, Ore.
THE QUAKING ASPENS WITH THEIR GREENISH WHITE BARK AND
FLUTTERING LEAVES RIVAL THE BIRCHES IN FRIENDLINESS
Groves like this are found in many forests, and they welcome the forest traveler

Once, riding in the rain on a gray October day, I came around a corner on the trail, and there, facing me, against a tall gray cliff, were two magnificent birches aflame with autumn foliage. It was as if suddenly, in that one spot, the sun was shining.

By the creek at the homeplace was an especially nice clump of birches with a few large flat rocks under them. It was a natural place for a special kind of garden. I planted lily of the valley and pheasant's-eye narcissus there, then, against the largest rock where I liked to sit and read, or sew or shell peas, or whatever I could do there, I planted columbine— wild red and yellow ones, and pale yellow alpine flowers.

I was at the homeplace a few days ago. Since I made the garden twenty years ago, the birches have grown tall. My little garden has spread along the creek bank. It's a lovelier place than ever, although it has had no care except for the shelter of the birches. Trees, left to themselves, seem able to look after their own.

Soon after we bought the homeplace, while we were still living in the old log house, we planted a golden willow in the yard. It was a good-sized tree in no time. I planted my first flower garden by it. The tree is huge now, standing alone in the field, for the old buildings are long gone. Seeing the tree a few days ago, I thought about my little garden. May I show it to you?

> I made a little garden,
> A lovely little garden,
> A fresh and fragrant garden
> Beneath my willow tree.
>
> I dug it and I hoed it,
> I watered it and sowed it,
> While all the little winds of heaven
> Came wandering by to see.

They whispered softly to it,
And lingered just to view it.
They breathed so gently through it,
And through the willow tree.

Flowers hung a welcome sign there,
And many came to dine there.
The butterfly sipped wine there—
The moth—the bumblebee.

The willow leaned above it,
The sky looked down to love it,
And singing birds made music
All day for thanks to me.

Because I made a garden,
A fragrant, lovely garden,
A very little garden
Beneath my willow tree.

After a forest fire that burned over half of the homeplace, a mountain maple came up on a bare hillside and grew prodigiously tall and wide. In fall it was a breathtaking sight, with color shading from yellow through orange to flame red. I suppose it is still there, but no longer visible from the house because the new growth of fir and tamarack has reached so tall.

Arrowwood and mountain maple may not deserve the name of "tree" since they aren't very tall, but, growing along a woods road or trail, reaching out into the open space for sunlight, the plumy flower clusters of arrowwood and the delicate tracery of maple boughs are among the fairest of all woods things.

There's another tree on the homeplace that is dear to the hearts of all the family. We planted a small, perfect fir by the door when we first built the new house. It was to be an outdoor Christmas tree. World War II came along and the REA line was postponed until after the war. When we finally got electricity and everyone was home again, the tree had grown quite large. We strung it with a hundred large colored lights. That night it snowed, covering every bough of the tree with a six-inch pillow of soft fluffy snow. The colored lights, shining through the snow, made our tree an unbelievable fairy creation. The weather turned cold and all through the holidays we had that wholly unexpected lovely thing.

That tree is taller than the house now, too big for the place where we planted it. The birds have taken it over. There are always nests in it. One summer cedar waxwings built a home and raised a family on a low branch just outside the window where we could watch everything they did. In winter chickadees take sunflower seeds there and hold them down with one foot while they open them. It's always full of birds waiting their turn at the feeder. Cut it down? Oh, not ever! Please—not ever!

A tree that has lived its life almost out often becomes an apartment house for wild things. On the homeplace there were a few such trees, tamaracks, remnants left after some forest fire of very many years ago, towering above the current crop of timber, showing cavities from an old burn about the base of them.

One such tree stood in our field for years. The top had broken off but the upper branches were still alive. From top to bottom it was drilled with many sizes of holes, all inhabited. The only pair of northern pileated woodpeckers in that area lived in it. In other holes were bluebirds, tree swallows, chickadees, little owls, flickers, and a family of pine squirrels. The flickers lived near the base of the tree. By tapping on the trunk they could set the tree to humming like the sound of a giant swarm of bumblebees. Finally, the old tree fell. There was a nest in every hole. There was still sound wood enough for many warm fires for our own home. A tree is useful to the very last scrap.

Once I stood on tiptoe to look into a used-looking hole in a tall stump. Out came a whole family of flying squirrels.

THE LONESOME PINE

Said to be the largest standing ponderosa pine in Idaho. Breast high, it measures slightly over six feet in diameter. Located only one mile from the tip of Beauty Bay, an arm of beautiful Coeur d'Alene Lake, this tree could tell many a tale. The Indians knew it for generations before Lewis and Clark made their westward journey. It was huge when the Jesuit missionary, Father DeSmet, built a log chapel entirely without nails. The chapel still stands at Cataldo, only a few miles from the Lonesome Pine.

Under loosened slabs of bark on another one, little brown bats slept.

A garden needs a tree, but trees are perfectly capable of maintaining their own gardens. In any forest, every tree will have its own complement of plants that thrive in its own particular kind of shade and ground cover. Under spruce trees are ground pine and sometimes Indian pipe. Fir trees, with tall, bare trunks, give little competition to underbrush, so we find bear grass, huckleberry bushes, paintbrush, red twinberries, squaw currant, wild asters, and many other plants. Pine woods are favored by grasses and small early spring flowers.

Poplar or quaking aspen grow along the edges of damp wild meadows. Tall shrubs such as red dogwood, hawthorn, black wineberry and wild rose make thickets under them. Birds like this combination of shelter, protection, and an ever-abundant food supply. You'll find many warblers, several sparrows, vireos, kingbirds, kinglets, hummingbirds and catbirds there. Have you heard the sweet, low, many-noted night song of the catbird? I hope you have.

Trees mean much to people. To wildlife they can mean life itself. A home without a tree lacks one of the important things it takes to make a home. And a country that hasn't kept its forests and appreciated them, and used them, is poor indeed.

APPENDIX

A

Accredited Schools of Forestry

Referring to Chapter 3

In 1966 the list of accredited forestry schools are:

Auburn University, Department of Forestry, Auburn, Alabama. 36830

University of California, School of Forestry, Berkeley, California. 94720

Clemson University, Department of Forestry. Clemson, South Carolina. 29631

Colorado State University, College of Forestry and Range Management, Fort Collins, Colorado. 80522

Duke University, School of Forestry (Graduate), Durham, North Carolina. 27706

University of Florida, School of Forestry, Gainesville, Florida. 32603

University of Georgia, School of Forestry, Athens, Georgia. 30601

University of Idaho, College of Forestry, Moscow, Idaho. 83843

University of Illinois, Department of Forestry, Urbana, Illinois. 61803

Iowa State University, Department of Forestry, Ames, Iowa. 50010

Louisiana State University, School of Forestry and Wildlife Management, Baton Rouge, Louisiana. 70803.

University of Maine, School of Forestry, Orono, Maine. 04473

University of Massachusetts, Department of Forestry and Wildlife Management, Amherst, Massachusetts. 01003

Michigan State University, Department of Forestry, East Lansing, Michigan. 48823

The University of Michigan, School of Natural Resources, Ann Arbor, Michigan. 48104

University of Minnesota, School of Forestry, St. Paul, Minnesota. 55101

University of Missouri, School of Forestry, Columbia, Missouri. 65201

Montana State University, School of Forestry, Missoula, Montana. 59801

University of New Hampshire, Department of Forestry, Durham, New Hampshire. 03824

North Carolina State of the University of North Carolina, School of Forestry, Raleigh, North Carolina. 27607

Oregon State University, School of Forestry, Corvallis, Oregon. 97331

The Pennsylvania State University, School of Forestry, University Park, Pennsylvania. 16802

Purdue University, Department of Forestry and Conservation, Lafayette, Indiana. 47907

State University College of Forestry at Syracuse University, Syracuse, New York. 13210

Utah State University, College of Forest, Range and Wildlife Management, Logan, Utah. 84321

University of Washington, College of Forestry, Seattle, Washington. 98105

West Virginia University, Division of Forestry, Morgantown, West Virginia. 26506

Yale University, School of Forestry (Graduate), New Haven, Connecticut. 06511

At least eighteen other institutions offer professional curricula in forestry. Some of them may have been added to the accredited list since this was prepared.

B

Forestry Curriculum

Referring to Chapter 3

In most Forestry Schools the Freshman students in all programs take the same subjects. These subjects might be:

General Forestry	Survey of the whole field of forestry—Classification, specialization, professional opportunities
Mathematics	
Chemistry	
English Composition	
General Botany	

Sophomore year (Forest Management)

Dendrology	Principal timber trees of United States—identification and characteristics
Mensuration	Measurement of standing and felled timber and timber products
Forest Protection	Major causes of forest damage including insects, disease and fire
Forest Engineering—Topographic Surveying	

Wood Technology	Wood structure, seasoning, grading, treatment. Wood identification
Forest Soils	Classification, relation to forest types, rate of growth and forest management
Plant Physiology	Concepts of plant growth
Basic Geology	Study of rocks and minerals
Principles of Economics	Money and banking, trade, taxation, labor, employment, business

Junior Year   (Forest Management)

Aerial Photo-interpretation	Techniques and principles of forest photo-interpretation. Typing and volume estimating from aerial photos
Mensuration-Timber Growth	Stands and individual trees
Forest Valuation	Appraisal of timber, land equipment and other assets
Forest Ecology	Environmental effect on occurrence and growth of forest vegetation
Silvicultural Practices	Treatment of stands to insure perpetuation of forest resources
Forestation	Growing and planting forest trees
Forest Recreation	Development of outdoor recreation. Policy and planning
Logging Methods	Relative merits of various methods
Forest Engineering	Public land survey. Triangulation. Drafting of field data
Abridged General Physics	Mechanics, heat, light, sound, electricity and magnetism

Senior year   (Forest Management)

Forest Economics	Credit, taxation, and marketing
Forest Administration	Administration and Personnel work of public and private organizations
Watershed Management	Integrated use for the production of water
Forest Management	Achieving and maintaining sustained yield
Fire Control	Fire control planning and administration
Range Management	Current developments, forage utilization, range condition
Wood Utilization	Wood-using industries, processes and products

Business Law   Conduct of business, proper legal prac-
                                     tices, contracts

Electives:

Some subjects are omitted such as Physical Education, Defense Training,
and English subjects.

C

University of Nebraska Alumni

Referring to Chapter 3

Dave Shoemaker was one of many graduates from the University of
Nebraska who became outstanding among the early leaders in conservation, both within and outside the Forest Service. Some of those who come
to mind are: W. R. Chapline, Chief of Range Research in the Forest Service for many years; C. L. Forsling, who became Assistant Chief of the
Forest Service in charge of research; Paul Roberts, who headed the Shelter
Belt Project and served as Assistant Regional Forester in charge of range
and wildlife management in the Northern Rocky Mountain Region of
the Forest Service; Lynn Douglas, who ended his career as Chief of Range
and Wildlife Management in the Pacific Northwest Region of the Forest
Service; Arthur Sampson, noted for basic range and watershed research as
first head of the Great Basin Experiment Station and head of the Range
Management Department for many years at the University of California;
L. J. Palmer, pioneer in reindeer research in Alaska; R. R. Hill, who did
range research at Fort Valley Experiment Station, served as Assistant
Chief of the Division of Range Management in Washington, D.C., and
was an Assistant Regional Forester in the Lake States Region; C. E.
Fleming, many years in charge of the Range Management Department at
the University of Nevada; Art Upson, a leader in tropical forestry; Carl
Kreuger, Assistant Regional Forester in charge of timber management at
Denver, Colorado; Fred D. Douthitt, Assistant Chief of the Division of
Range Management in the California Region; Leon Hurtt, active in range
research in Montana; G. B. McDonald, well-known dean of the College
of Forestry at Iowa State University; R. P. Wollenberg, prominent in
private forestry in the Pacific Northwest; the three Benedict brothers—
one an early inspector out of the Washington office, Miller S., a longtime
Forest Supervisor in the Intermountain Region, and Maurice A., a prominent Forest Supervisor in the California Region of the Forest Service;
Ralph Bodily, Ed Polson, L. L. Bishop, Carlos Bates, and doubtless others
equally distinguished. All of these men filled important assignments in
addition to those mentioned.

D

Loggers' Lingo

Referring to Chapter 6

Bindle stiff: one who works a short while, draws his pay, and heads for the nearest town to blow his money on women and booze.

Bull cook *or* Crumb boss: one who built fires, carried wood—a chore boy.

Bull of the woods: superintendent.

Catawampus: out of line.

Choker setter: the man who hitched the cable to the logs to be skidded to the mill.

Dehorn: any kind of booze.

Gandy dancer: logging railroad section hand.

Grease monkey: man who swabbed or spread grease on log chutes to reduce the friction of the logs.

Gut robber: an inferior or overly economical cook.

Hardtack outfit: an outfit providing poor meals.

Haywire: applied to anything below standard or "not up to snuff."

Highball outfit: hell-roarin' outfit, always at high speed; in a rush.

Idiot stick: shovel.

Inkslinger: a timekeeper.

Iron burner: a blacksmith.

Macaroni: sawdust.

Misery whip: crosscut saw.

Molly Hogan: a temporary repair, a patch-up job.

Muzzle loader: bunk beds side by side, where you crawled in over the end.

Powder monkey: one who carries on blasting operations.

Scandinavian dynamite: snoose, Swedish conditioning powder.

Skid row: a pole road over which logs were pulled by oxen. Also a part of the city where loggers congregated; not now limited to loggers.

Swedish fiddle: crosscut saw.

"Timber!": warning cry when a big tree is about to fall.

Town clown: small-town policeman.

Widowmaker: a dangerous limb or snag hanging in or falling from a tree.

E

NATIONAL FOREST REGIONS

Chief, Forest Service

Referring to Chapter 13

U.S. Department of Agriculture
South Building, Twelfth and Independence Ave. S.W.
Washington, D.C.   20250

Region 1

Northern Region
Federal Building, Missoula, Montana   59801

National Forests

Montana

Beaverhead	State Highway 41 & Skihi St., Dillon 59725
Bitterroot	316 N. Third St., Hamilton 59840
Custer	1015 Broadwater, Billings 59101
Deerlodge	107 E. Granite, Butte 59701
Flathead	290 N. Main, Kalispell 59901
Gallatin	The Story Building, 37 Main St., Bozeman 59715
Helena	616-618 Helena Ave., Helena 59601
Kootenai	Libby 59923
Lewis and Clark	Federal Building, Great Falls 59401
Lolo	Federal Building, Missoula 59801

Idaho

Clearwater	360 Michigan Ave., Orofino 83544
Coeur d'Alene	218 N. Twenty-third St., Coeur d'Alene 83814
Kaniksu	Sandpoint 83864
Nezperce	Mill & Main Sts., Grangeville 83530
St. Joe	St. Maries 83861

Washington

Colville	Colville 99114

Region 2

Rocky Mountain Region
Denver Federal Center Building,
85 Denver, Colorado 80225

National Forests

Colorado

Arapaho	1010 Tenth St., Golden 80402
Grand Mesa–Uncompahgre	Postoffice Building, Delta 81416
Gunnison	216 Colorado, Gunnison 81230
Pike	403 S. Cascade, Colorado Springs 80901
Rio Grande	Fasset Building, Monte Vista 81144
Roosevelt	Postoffice Building, Fort Collins 80522
Routt	Hunt Building, Steamboat Springs 80477
San Isabel	Postoffice Building, Pueblo 81002
San Juan	West Building, Durango 81301
White River	Postoffice Building, Glenwood Springs 81601

South Dakota

Black Hills	Forest Service Building, Custer 57730

Nebraska

Central Plains Forestry Office	Postoffice Building, Lincoln 68508

Wyoming

Bighorn	Columbus Building, Sheridan 82801
Medicine Bow	1948 Grand Ave., Laramie 82071
Shoshone	Blair Building No. 1, Cody 82414

Region 3

Southwestern Region
Federal Building, 517 Gold Ave. S.W.
Albuquerque, New Mexico 87101

National Forests

Arizona

Apache	Postoffice Building, Springerville 85938

Coconino	114 N. San Francisco St., Flagstaff 86002
Coronado	Postoffice Building, Tucson 85702
Kaibab	107 N. Second St., Williams 86046
Prescott	344 S. Cortez, Prescott 86301
Sitgreaves	113 W. Hopi Dr., Holbrook 86025
Tonto	Room 6208 Federal Building, Phoenix 85025

New Mexico

Carson	Forest Service Building, Taos 87571
Cibola	304 Courthouse Building, Albuquerque 87103
Gila	301 W. College Ave., Silver City 88061
Lincoln	Federal Building, Alamogordo 88310
Santa Fe	Federal Building, Santa Fe 87501

Region 4

Intermountain Region
Federal Building, Ogden, Utah 84403

National Forests

Idaho

Boise	413 Idaho St., Boise 83702
Caribou	427 N. Sixth Ave., Pocatello 83201
Challis	Forest Service Building, Challis 83226
Payette	Forest Service Building, McCall 83638
Salmon	Forest Service Building, Salmon 83467
Sawtooth	1525 Addison Ave. E., Twin Falls 83301
Targhee	48 E. First North, St. Anthony 83445

Nevada

Humboldt	Postoffice Building, Elko 89801
Toiyabe	1555 S. Wells Ave., Reno 89502

Utah

Ashley	Postoffice Building, Vernal 84078
Cache	429 S. Main, Logan 84321
Dixie	500 S. Main St., Cedar City 84720
Fishlake	170 N. Main, Richfield 84701

Manti–La Sal	350 E. Main St., Price 84501
Uinta	Federal Building, Provo 84601
Wasatch	4438 Federal Building, 125 S. State, Salt Lake City 84111

Wyoming

Bridger	Forest Service Building, Kemmerer 83101
Teton	Forest Service Building, Jackson 83001

Region 5

California Region
630 Sansome St., San Francisco, California 94111

National Forests

California

Angeles	1015 N. Lake St., Pasadena 91104
Cleveland	Room 209 Russ Building, San Diego 92101
Eldorado	Placerville 95667
Inyo	207 W. South St., Bishop 93514
Klamath	1215 S. Main St., Yreka 96097
Lassen	707 Nevada St., Susanville 96130
Los Padres	Federal Building, Santa Barbara 93101
Mendocino	Willows 95988
Modoc	Alturas 96101
Plumas	Quincy 95971
San Bernardino	Civic Center Building, San Bernardino 92401
Sequoia	P.O. Box 391, Porterville 93258
Shasta-Trinity	1615 Continental St., Redding 96001
Sierra	4831 E. Shields Ave., Fresno 93726
Six Rivers	710 E. St., Eureka 95501
Stanislaus	175 S. Fairview Lane, Sonora 95370
Tahoe	Highway 49, Nevada City 95959

Region 6

Pacific Northwest Region
729 N.E. Oregon St., P.O. Box 3623, Portland, Oregon　97208

National Forests

Oregon

Deschutes	745 Bond St., Bend　97701
Fremont	Center & G Sts., Lakeview　97630
Malheur	139 N.E. Dayton St., John Day 97845
Mount Hood	P.O. Box 5241, Portland　97216
Ochoco	Bottero Building, Prineville　97754
Rogue River	Postoffice Building, Medford　97501
Siskiyou	1504 N.W. Sixth St., Grants Pass 97526
Siuslaw	545 S. Second St., Corvallis　97330
Umatilla	116 S. Main St., Pendleton　97801
Umpqua	Federal Office Building, Roseburg 97470
Wallowa-Whitman	Main & Auburn Sts., Baker—97814
Willamette	210 E. Eleventh St., Eugene—97401
Winema	411 Main St., Klamath Falls　97601

Washington

Gifford Pinchot	1408 Franklin St., Vancouver　98660
Mount Baker	Federal Office Building, Bellingham 98225
Okanogan	Postoffice Building, Okanogan　98840
Olympic	Postoffice Building, Olympia　98502
Snoqualmie	905 Second Avenue Building, Seattle 98104
Wenatchee	3 S. Wenatchee Avenue, Wenatchee 98801

Region 7

Eastern Region
6816 Market St., Upper Darby, Pennsylvania　19082

National Forests

Kentucky

Cumberland	Postoffice Building, Winchester　40391

New Hampshire and Maine

White Mountain   Federal Building, 719 Main St., Laconia, New Hampshire 03246

Pennsylvania

Allegheny   Postoffice Building, Warren 16356

Vermont

Green Mountain   22 Evelyn St., Rutland 05702

Virginia

George Washington   Federal Building, Harrisonburg 22801
Jefferson   Carlton Terrace Building, Roanoke 24001

West Virginia

Monongahela   Department of Agriculture Building, Sycamore St., Elkins 26241

Region 8

Southern Region
50 Seventh St. N.E., Atlanta, Georgia 30323

National Forests

Alabama

National Forests in Alabama   502 Washington St., Montgomery 36101
William B. Bankhead
Conecuh
Talladega
Tuskegee

Arkansas

Ouachita   Box 1270, Hot Springs National Park 71902

Ozark and St. Francis   Russellville 72801

Florida

National Forests in Florida   214 S. Bronough St., Box 1050, Tallahassee 32302
Apalachicola
Ocala
Osceola

Georgia

National Forests
  in Georgia
Chattahoochee
Oconee

322 Oak St., N.W. Box 643,
  Gainesville 30501

Louisiana

Kisatchie

Building 6, Veterans Hospital,
  Box 471, Alexandria 71302

Mississippi

National Forests
  in Mississippi
Bienville
Delta
De Soto
Holly Springs
Homochitto
Tombigbee

380 Milner Building,
  Box 1291, Jackson 39205

North Carolina

National Forests in
  North Carolina
Croatan
Nantahala
Pisgah
Uwharrie

50 S. French Broad,
  Box 731, Asheville 28802

South Carolina

National Forests in
  South Carolina
Francis Marion
Sumter

Room 601A, Federal Building,
  Columbia 29201

Tennessee

Cherokee

Federal Building, Box 400, Cleveland
  37312

Texas

National Forests
  in Texas
Angelina
Davy Crockett
Sabine
Sam Houston

307 S. First St., Box 969,
  Lufkin 75902

Region 9

North Central Region
710 N. Sixth St., Milwaukee, Wisconsin 53203

National Forests

Illinois

Shawnee — Harrisburg National Bank Building, Harrisburg 62946

Indiana and Ohio

Wayne-Hoosier — Stone City National Bank Building, Bedford, Indiana 47421

Michigan

Hiawatha — Postoffice Building, Escanaba 49829
Huron-Manistee — Cadillac 49601
Ottawa — Ironwood 49938

Minnesota

Chippewa — Cass Lake 56633
Superior — Duluth 55801

Missouri

Clark — Rolla 65401
Mark Twain — 304 Wilhoit Building, Springfield 65806

Wisconsin

Chequamegon — Federal Building, Park Falls 54552
Nicolet — Merchants State Bank Building, Rhinelander 54501

Region 10

Alaska Region
Fifth Street Office Building, Box 1631, Juneau 99801

National Forests

Chugach — 328 E. Fourth Ave., Anchorage 99501
North Tongass — 217 Second St., Juneau 99801

INDEX